SpringerBriefs in Applied Sciences and Technology

SpringerBriefs present concise summaries of cutting-edge research and practical applications across a wide spectrum of fields. Featuring compact volumes of 50–125 pages, the series covers a range of content from professional to academic.

Typical publications can be:

- A timely report of state-of-the art methods
- An introduction to or a manual for the application of mathematical or computer techniques
- A bridge between new research results, as published in journal articles
- A snapshot of a hot or emerging topic
- An in-depth case study
- A presentation of core concepts that students must understand in order to make independent contributions

SpringerBriefs are characterized by fast, global electronic dissemination, standard publishing contracts, standardized manuscript preparation and formatting guidelines, and expedited production schedules.

On the one hand, **SpringerBriefs in Applied Sciences and Technology** are devoted to the publication of fundamentals and applications within the different classical engineering disciplines as well as in interdisciplinary fields that recently emerged between these areas. On the other hand, as the boundary separating fundamental research and applied technology is more and more dissolving, this series is particularly open to trans-disciplinary topics between fundamental science and engineering.

Indexed by EI-Compendex, SCOPUS and Springerlink.

More information about this series at http://www.springer.com/series/8884

Rabiu Muazu Musa · Anwar P. P. Abdul Majeed ·
Norlaila Azura Kosni · Mohamad Razali Abdullah

Machine Learning in Team Sports

Performance Analysis and Talent Identification in Beach Soccer & Sepak-takraw

Springer

Rabiu Muazu Musa
Department of Credited Co-curriculum
Centre for Fundamental
and Continuing Education
Universiti Malaysia Terengganu
Kuala Terengganu, Malaysia

Norlaila Azura Kosni
Faculty of Sport Science and Recreation
Universiti Teknologi MARA
Bandar Tun Abdul Razak Jengka, Pahang, Malaysia

Anwar P. P. Abdul Majeed
Innovative Manufacturing,
Mechatronics and Sports Laboratory
Faculty of Manufacturing
and Mechatronics
Engineering Technology
Universiti Malaysia Pahang
Pekan, Pahang, Malaysia

Mohamad Razali Abdullah
East Coast Environmental
Research Institute
Universiti Sultan Zainal Abidin
Kuala Terengganu, Malaysia

ISSN 2191-530X   ISSN 2191-5318   (electronic)
SpringerBriefs in Applied Sciences and Technology
ISBN 978-981-15-3218-4   ISBN 978-981-15-3219-1   (eBook)
https://doi.org/10.1007/978-981-15-3219-1

This Springer imprint is published by the registered company Springer Nature Singapore Pte Ltd.
The registered company address is: 152 Beach Road, #21-01/04 Gateway East, Singapore 189721, Singapore

This book is dedicated to my beloved family Late Malam Muazu Musa, Halima Muazu, Adamu Muazu, Sadiya Muazu and Abubakar Muazu as well as my precious wife Nafisa Usman Yahaya.

Rabiu Muazu Musa

I would like to dedicate this book to my family, P. P. Abdul Majeed K. Idros, Sulleha Eramu, Noriza P. P. Abdul Majeed, as well as my wife Sharifah Maszura Syed Mohsin and not forgetting the apple of my eyes, Saffiya Anwar.

Anwar P. P. Abdul Majeed

This book is wholeheartedly and humbly dedicated to my beloved parents, Kosni Awang and Khayati Hamzah as well as my dearest husband, Ahmad Ahsan Syarafi and daughter, Annaila Amanda Safiyyah.

Norlaila Azura Kosni

I would like to dedicate this book to my wife, my daughter and my son.

Mohamad Razali Abdullah

Acknowledgements

We would like to acknowledge Prof. Dr. Zahari Taha for providing us with the necessary knowledge, guidance and valuable suggestions for making the realisation of the book possible. We would also like to extend our gratitude to the 2017 Terengganu Asian Beach Soccer Committee (AFC 2017-TSG). We would also wish to acknowledge the coaches as well as the managers of the Malaysian sports schools (Sekolah Sukan Tunku Mahkota Ismail and Sekolah Sukan Malaysia Terengganu) for their support in the accomplishment of this project.

Rabiu Muazu Musa
Anwar P. P. Abdul Majeed
Norlaila Azura Kosni
Mohamad Razali Abdullah

Contents

Chapter 1
An Overview of Beach Soccer, Sepak Takraw and the Application of Machine Learning in Team Sports

1.1 An Overview of Beach Soccer Sport

Beach soccer is considered amongst the world's most rapidly growing sports, especially since the recognition of the sport by FIFA and its subsequent promotion as well as organisations into various competitive matches, seminars and other initiatives [1]. The sport is normally played on a sandy surface amounting to a depth of about 40 cm. Similar to the nature of futsal and basketball, the beach soccer match is separated into three periods of 12 min with a resting time of 3 min between the matches. It is important to note that the time is stopped when the play is interrupted and as such, the total duration of the match runs for about 36 min. The match is temporarily stopped during an emergency for doctor assistance or awarding for a penalty. A team for beach soccer is made up of ten players; five players on the field and the remaining five are sidelined for reserved purposes. It is worth noting that no restriction on the number of substitutions; therefore, a continuous change of players is allowed as the match progresses for the purpose of maintaining tempo as well as intensity during the match. A referee, as well as two assistant referees (linesmen), is normally saddled with the responsibility of officiating the match.

The first organised FIFA Beach Soccer World Cup was held in Brazil in 2005, a year after the sport was consolidated into the FIFA organisation. The final match of the world cup was played between France and Portugal which attracted many spectators and grasped the attention of various nations worldwide [2]. Owing to the success of the said world cup, the sport rapidly advanced and a number of researchers, as well as other relevant stakeholders, endeavour to look into the nature of the sport and the factors that could influence its performance with a view to providing data that could be used to guide performance and ensure success within the sport.

© The Author(s), under exclusive license to Springer Nature Singapore Pte Ltd. 2020
R. Muazu Musa et al., *Machine Learning in Team Sports*,
SpringerBriefs in Applied Sciences and Technology,
https://doi.org/10.1007/978-981-15-3219-1_1

1.2 An Overview of Sepak Takraw Sport

The sport of sepak takraw was originated from a traditional game that was commonly popular within the South Asian countries. It is worth noting that the sport was initiated and began in Malaysia, and therefore, the game is considered as a traditional Malaysian sport before it began spreading into the other South Asian countries. Initially, the sport was mainly played based on the rules set by the players. The size of the court, the duration of play as well as the winning of a set are largely decided by the team involved in the game. The nature of the sport constitutes the combined elements of soccer, volleyball, basketball, badminton, gymnastics as well as the ancient sport of sepak raga. Owing to the rapid advancement as well as the wider recognition of the sport worldwide, the game was modified to a high-performance sport and a standardised set of rules was introduced to govern the match play [3].

The introduction of the sepak takraw sport in the 10th Asian Games in Beijing in 1990, as well as the subsequent presentation of the sport for demonstration in 1998 Commonwealth Games held in Kuala Lumpur, Malaysia, has further contributed to the heightened popularity of the sport. It has been reported that the sport is one of the fastest-growing in Asia, and more than 20 countries worldwide encompassing Argentina, Australia, Brazil, Canada, Korea, Germany, England, India, Japan, Puerto Rico, Spain and the USA have already embraced the sport [4]. The game of sepak takraw is played on an area size of twice the badminton court. The court is separated by a net of a 1.52 m high. A team is made up of three players that are commonly known as the feeder, 'tekong'/server as well as spiker or killer.

1.3 Machine Learning in Team Sports

Owing to the advancement of computational intelligence, machine learning models have been employed in order to identify or predict different performance aspects in team sports. It is worth noting that, more often than not, most studies are focused on predicting the outcome of the match of a given team sport, for instance, football, cricket, baseball and basketball to name a few. Ulmer and Fernandez evaluated different classifiers, i.e. baseline, Gaussian Naive Bayes (GNB), multinomial Naive Bayes, hidden Markov model, linear support vector machine (SVM), radial basis function (RBF) SVM, and random forest (RF), in predicting win, draw or lose of English Primer League (EPL) for ten seasons (2002/2003–2011/2012) [5]. A number of features were selected, viz. home team, away team, score, winner and the number of goals for each team. Nonetheless, it is worth to mention that the findings of the investigation were somewhat inconclusive, as the models evaluated could not attain relatively good classification accuracy (CA). This may be attributed owing to the arbitrary nature and the variety of the tactical and technical capabilities of the teams over the decade.

The employment of artificial neural network (ANN) and logistic regression (LR) has also been investigated in predicting EPL match outcomes [6]. The authors utilised home and away goals, home and away shots, home and away corner, home and away odds, home and away attack strength, home and away players' performance index, home and away managers' performance index, home and away managers' win, home and away streak as the features to the models developed. A total of 110 matches played in the 2014/2015 EPL season was used for the study, in which only 20 sets of matches were used as the test set, whilst the rest are used as the train set. It was demonstrated from the study that the ANN model is able to achieve a CA of 85%, whilst the LR yielded 93% accuracy. However, it should be pointed out that a direct comparison of the models developed should not have been made, as the ANN model classified three classes, viz. win, lose or draw, whilst the LR model only classifies win or loss. The authors extended the study in [7], by exploring the efficacy of a Gaussian-based SVM model in predicting match outcomes utilising the same features. Nevertheless, only a CA of 53.3% was achieved by the model in classifying correctly the win, lose or draw.

A polynomial classifier was developed to classify the match outcomes of different leagues, i.e. EPL, season 2014/2015; La Liga Primera Division (LLPD), season 2014/2015; and Brazilian League Championships, seasons 2010 (BLC 2010) and 2012 (BLC 2012) [8]. A total of 19 features were selected for the EPL and LLPD, whilst 54 features were selected for the BLC. The proposed model was evaluated against different ML models available in WEKA, viz. naïve Bayes (NB), decision tree (DT), multi-layer perceptron (MLP), radial basis function (RBF) and support vector machine (SVM). It was shown that the proposed algorithm managed to attain an accuracy of 0.99, 0.99, 0.99 and 0.98, for EPL, LLPD, BLC 2010 and BLC 2012, respectively.

Joseph, Fenton and Neil investigated the efficacy of an expert Bayesian Network (eBN) against decision tree learner (MC4), NB, data-driven Bayesian, and a k-nearest neighbour (k-NN) classifier in predicting the match outcome for Tottenham Hotspur Football Club (1995/1996 and 1996/1997 seasons) [9]. The features considered are the availability of the key players evaluated at that time, the playing of a particular player in midfield, the quality of the opposing team, the quality of the Spurs attack, the quality and performance of the Spurs' team as well as the venue, i.e. home or away. It was shown from the investigation that the eBN provided a better CA of 59.21% against other classifiers evaluated.

Razali et al. [10] utilised BN to predict EPL match results for seasons between 2010 and 2013. The features selected are similar to of [8] with the exclusion of home and away team hit woodwork, home and away team offside as well as the inclusion of half-time home and away team goals and full-time home and away team goals. The tenfold cross-validation was used for the entire dataset, and the BN classifier was reported to be able to provide a CA of 75.09%.

The prediction of the match outcomes (win, lose or draw) of the top six EPL clubs (Manchester United, Manchester City, Liverpool, Arsenal, Chelsea, and Tottenham Hotspur) was investigated in [11]. The data was extracted over the span of four seasons (2013–2016). Fifteen features were selected based on their relevance to the game. Different variations of SVM models, namely linear, quadratic, cubic, fine radial basis function (RBF), medium RBF, as well as course RBF were developed to predict the match outcomes. A fivefold cross-validation technique was employed on the train data, whilst a separate fresh data was supplied to the best-trained model in order to evaluate its predictive efficacy. It was reported that the linear SVM model demonstrated an excellent CA of 100% on both the trained and test data.

Saricaoğlu et al. employed different machine learning models in order to predict the outcome of the Turkish Super League (TSL) [12]. The data used for the development of the prediction models is from 2013 to 2018, in which a train-to-test ratio of 80:20 is used. The features used were the team power ration ratio, exponential moving average performance points of teams based on the last eight match results of a given team, home and away performance, league point difference as well as the number of the home team audience. The data was normalised prior to developing the models. The ML models evaluated were LDA, QDA, k-NN, LR, DT, bagging as well as SVM. In addition to the aforesaid models, an ensemble model based on the models was developed. It was demonstrated from the study that the ensemble method provided the overall best match prediction with a classification accuracy of 62.03%.

Chai et al. investigated the efficacy of hybrid ensemble methods towards basketball outcome prediction [13]. The data was attained from the official Chinese Basketball Association website for the 2016/2017 regular season that consists of 20 teams, in which 210 games were used as a training set whilst the 70 games are used as the test set. The features selected were two-point, three-point, free-throw, attack, defence, assist, foul, steal, blunder, blocked-shot, fast break and dunk. A weighted combination scheme was utilised in order to identify the significant features that contribute towards the classification accuracy. Different SVM models (varied with respect to its kernels, i.e. linear, polynomial, sigmoid and RBF) were compared with other conventional models, namely LR, ANN and NB, and it was shown that the RBF-based SVM model outperforms the rest. Subsequently, different ensemble strategies, namely bagging and AdaBoost, are investigated. Furthermore, the use of random subspace method is also incorporated to develop the proposed framework. It was shown from the study that the proposed framework outperforms other models investigated with a classification accuracy of 84% and an $F1$ score of 82%.

Different ensemble learning models, particularly RF and XGBoost, have been investigated in predicting the shooting success of National Basketball Association (NBA) players [14]. The dataset utilised in the study is shots obtained from 2014/2015 NBA regular season. The features used are shot clock, final margin, period, dribbles, touch time, shot distance, points type, points, front guard and close defender's distance. Both models were built based on the shot clock, dribbles, shot distance and close defender's distance. It is worth noting that the 70:30 train-to-test ratio was used in the study. It was shown from the study that the XGBoost model with hyperparameter tuning yields the best shot prediction with an accuracy of 68%.

The classification of the win–loss outcome of Major League Baseball (MBL) by means of different machine learning models was investigated by Valero [15]. Sixty features were used to develop k-NN, MLP, SVM and DT models extracted between the years 2005 and 2014. The tenfold cross-validation technique was employed in training the models, and the efficacy of the models in classifying win or loss was evaluated via the CA. The WEKA Environment was used in developing the models. It was shown from the study that the SVM model was able to classify accurately the match outcomes to up to 59% against other models evaluated.

Tolbert and Trafalis investigated different variations of SVM models in predicting the match outcomes of MLB [16]. Fourteen different features were used to develop classifiers. The SVM classifiers were varied based on the kernels, in which linear, quadratic, cubic and RBF kernels were investigated. The tenfold cross-validation technique was employed in training the model. It was demonstrated from the study that the linear variation of the SVM classifier was able to predict accurately the winner and loser for the 2015 World Series.

ML has also been applied to predict the outcome of cricket matches. Pathak and Wadhwa employed different ML models, namely NB, SVM and RF to predict the outcome of One Day International (ODI) cricket match [17]. The features utilised were tossing outcome, home game advantage, day/night effect and bat first. The 80:20 train-to-test ratio was applied to the dataset. The balanced accuracy, as well as the Kappa statistic, was used to evaluate the efficacy of the models developed. It was established from the study that the NB model is suitable in predicting the match outcomes, owing to its ability to handle imbalanced datasets as remarked by the authors.

Kumar et al. compared the efficacy of MLP and DT in predicting the ODI match outcome [18]. The features utilised in their investigation were the team's past performance, ground, innings and home game advantage. The dataset comprises a total of 7494 ODI match records (from 1971 to 2017), and approximately 75% of the dataset was used as the training dataset, whilst the remaining was used as the test dataset. It was shown from the study that the MLP classifier (57.4%) provided a better albeit marginal predictive capability in comparison to the DT classifier (55.1%).

The match outcome prediction of the Bangladesh team in ODI was investigated in [19]. A number of classifiers developed in WEKA were tested, i.e. NB, SVM, DT, RF and k-NN. The data was extracted from ODI matches from the year 2005 to 2017. A total of seventeen features were used for the development of the classifiers. It was demonstrated from the study that the DT model is able to provide reasonable accuracy level in predicting the match outcomes. It is therefore evident from the selected literature reviewed, albeit non-exhaustive in this subsection somewhat suggests the capability of ML providing certain insight towards the performance of a given sport.

1.4 Principal Component Analysis

Principal component analysis (PCA) is a statistical technique that is often employed to recognise the pattern of a dataset from an observed group of variables [20]. PCA provides insights on the important variables that could represent a given dataset through considering the spatial as well as the temporal variability of an entire dataset.

The process of extracting information from the PCA is performed by excluding the less important component from a given data and consequently retaining the most valuable information from the data [21, 22]. The usage of PCA is non-trivial in extracting the most vital information from a large volume of a dataset which could assist in saving effort, cost as well as time since the actual information is normally retained.

1.5   Cluster Analysis

1.5.1   Hierarchical Agglomerative Cluster Analysis (HACA)

Hierarchical agglomerative cluster analysis (HACA) is considered as an exploratory as well as an unsupervised method in which a hierarchy of clusters is developed with regard to a particular observation and consequently a number of similar observations are formed into a distinct observation [23]. It is worth to mention that the learning process in this algorithm is determined by the merges as well as the splits of a dataset which often took place in a greedy manner such that similar observations are segregated and demonstrated in a dendrogram [24]. It should be noted that in HACA, the number of clusters is displayed by the dendrogram based on the proximity ascertained by a given or predetermined clusters. The cosine's distance was utilised in this study, and the validation technique of the clustering was carried out by means of class centroids [25].

1.5.2   Louvain Clustering

The Louvain clustering algorithm is often seen as the state-of-the-art-based clustering algorithm which clustered a given data in two distinctive steps; in the first step, the algorithm looks for a 'small' community through optimising the modularity in a conventional technique. In the second step, the algorithm aggregates nodes of similar communities and subsequently forms a distinctive community that constitutes a new network whose nodes are the communities [26]. These processes are carried out iteratively until a threshold of modularity is achieved. It is worthy to note that this step often leads to a hierarchical decomposition of the network and as such several partitions are formed [27]. The partitions are normally the density of edges within the community as opposed to the edges of the inter-communities.

1.6   Classification Algorithms

1.6.1   k-Nearest Neighbour

Dubbed as a 'lazy learning' algorithm, k-nearest neighbour (k-NN) is one of the simplest yet powerful types of supervised ML algorithms. As opposed to eager learning, lazy learning suggests that no generalisation is made on the data points. k-NN is also known to be a non-parametric (assumptions are not made from the data)-based model that has been extensively used in both classifications [28, 29] and regression [30, 31] problems, albeit it was initially [32] introduced for classification. The decision boundary of a class is determined by the number of neighbours (viz. k) that surrounds a point of interest and the distance (often Euclidean, but other distance metrics are also used, for instance, Manhattan, Hamming and Minkowski amongst others) between the points with the k. Nonetheless, it is worth noting that the selection of k is non-trivial as a small number of k would provide a low bias, but high variance fit, whilst large k values are susceptible to a lower variance, but an increased bias.

1.6.2   Support Vector Machine

Support vector machine (SVM) is an ML model developed by Cortes and Vapnik [33] that discriminates between classes through the use of hyperplane with a maximum margin. For a linearly separable data in a two-dimensional space, the hyperplane is simply a line dividing the plane accordingly with respect to its classes. Nonetheless, more often than not, the dispersion of the data is non-linear, a transformation is required, and this is known as the 'kernel trick'. The kernel function (polynomial or Gaussian radial basis function) transforms the data into higher-dimensional feature space to allow for a linear separation to be possible. The margin width is governed by the regularisation hyperparameter, C. A small value of C will increase the width of the margin, whilst a large value of C will conversely decrease the margin width. Another important hyperparameter for non-linear-based kernels is the gamma (γ) parameter. This parameter essentially influences the spread of the decision region. A low γ value will consider points that are far from the separation line and hence borders the decision region. Conversely, a high γ value considers points that are close to the line, consequently tightens the decision boundary.

1.6.3   Artificial Neural Networks

McCulloch and Pitts [34] in the 1940s brought about the fundamental notion of what we know now as artificial neural network (ANN) that is inspired by how the brain works and has been utilised in a myriad of classification and regression problems [35–38]. ANN comprises multiple layers, i.e. the input layer, hidden layer(s) as well as the output layer, and hence, it is often also referred to as multi-layer perceptron (MLP). The network becomes deep by increasing the number of hidden layers. The hidden layers are also known as the 'distillation layer' as it distils (extracts) significant features to the subsequent layer. The input layer consists of input nodes (neurons). The common activation functions that transform the sum of weighted inputs to obtain an output are rectified linear unit (ReLu), sigmoid, tanh and linear amongst others. The unique architecture of the ANN allows it to model complex non-linear relationships and yield exceptional predictive capability.

1.6.4 Random Forests

Random forests (RF) is an ensemble algorithm that consists of a set of decision trees (DT), in which the best solution is attained via voting based on the predictive efficacy of each DT [39]. Through this method of bootstrap aggregating (also known as bagging), it reduces the effect of noise, hence mitigating the notion of overfitting. The DT is generated based on a randomly split dataset. The Gini importance is utilised to evaluate the importance of the features and hence monitors the decrease in node impurity. The dataset that was not used for developing the models is known as the 'Out-Of-Bag' sample, and it is used to evaluate the efficacy of the developed models.

1.7 Study Participants

1.7.1 Beach Soccer

The participants of the beach soccer sport in this study consist of all the teams that were involved in the AFC Beach Soccer Tournament 2017. A sum of 12 teams took part in the tournament which comprised of China PR, Bahrain, Afghanistan, Malaysia, Oman, Lebanon, Thailand, Japan, United Arab Emirates (UAE), Iraq, Islamic Republic of Iran (I.R. Iran) as well as Qatar. The overall performances of the teams involved were notated and analysed throughout the competition period. A total number of 23 matches from the 12 teams were analysed. It is worth to mention that prior to the commencement of data collection, all the coaches, managers as well as the organising committee were informed about the aim of the study and verbal consent was obtained via Terengganu State 2017 Beach Soccer Technical Committee (AFC2017-TSG).

1.7.2 Sepak Takraw

The participants of the sepak takraw sport in the current study consist of a total number of 74 youth players with mean age and standard deviation (15.1, 1.32), respectively. The players were recruited from two sports schools in Malaysia (Sekolah Sukan Tunku and Sekolah Sukan Malaysia Terengganu). Before the commencement of any data acquisition in the present study, all the coaches, managers, as well as relevant stakeholders, were informed about the purpose of the research and informed consent was obtained from all the participants.

1.8 Performance Analysis

1.8.1 Beach Soccer Analysis

A total number of 20 technical as well as tactical performance indicators were established. The performance indicators were utilised to notate the performances of the teams. It is essential to note that four professional coaches who possessed an average of 10 years of coaching experience validated the performance indicators developed. The StatWatch application, an android-based application for notational analysis was used as a tool for the analysis of the teams' performances in accordance with the procedures previously documented by the precedent investigators [40]. Two independent performance analysts were instructed to notate the performance of the teams such that each analyst was responsible for covering a particular team at a time. It is worthy to mention that prior to the full analysis, the analysts were familiarised with the performance indicators selected. Reliability analysis was carried out using footage from a different match. For the purpose of establishing the consistency as well as to test the observational errors on the performance indicators developed, the performance analysts were instructed to notate the match individually, and their agreement was subsequently compared. Cohen's Kappa statistical test and Cronbach's alpha analysis were employed to measure the agreement and the consistencies of the analysts with regard to the performance indicators [41]. It is non-trivial to note that a Kappa's value of 0.94, as well as a Cronbach's alpha of 0.98, was observed indicating a strong agreement and consistency amongst the performance analysts in their overall analysis.

1.8.2 Sepak Takraw Analysis

A sum of 22 performance indicators was developed for the analysis of the sepak takraw players. These performance indicators which include service ace (SA), service effective (SE), service weak (SW), service net (SN), service out (SO), break ball success (BrS), break ball weak (BrW), break ball fail (BrF), passing success (PS), passing weak (PW), passing fail (PF), passing desperate (PD), blocking success (BS), blocking weak (BW), blocking fail (BF), killing ace (KA), killing effective (KE), killing weak (KW), killing net (KN), killing out (KO), foul (F) and opponent mistake (OM) were considered based on their relevance to the sport of sepak takraw. A high-resolution camera was used to record all the matches during the national youth circuit league. The camera was suitably placed beside the court where a wider angle could be fully captured from the top of the court. The distance from the sideline of the court as well as the height was about 15 and 5 metres, respectively. A total of 217 national age group matches were analysed. The percentage performance of each player was used further for statistical analysis.

For the purpose of reliability testing, a number of 12 experienced performance analysts who were further trained and acquainted with the performance indicators selected were responsible for notating the match recorded from a wider screen projector. It is worth to mention that prior to the commencement of the full analysis, a random video match of sepak takraw was selected and all the performance analysts were requested to notate the performance of a particular player using the aforesaid performance indicators. Reliability analysis was computed by means of Cohen's Kappa inter-tester reliability testing to determine whether the performance analysts agreed unanimously on the actions performed by the player. A noteworthy agreement between the analysts was observed ($K = 0.76$, $p < 0.001$), which reflected that the agreement between the analysts was beyond a chance.

References

1. J. Castellano, D. Casamichana, Heart rate and motion analysis by GPS in beach soccer. J. Sports Sci. Med. **9**, 98–103 (2010). http://www.doaj.org/doaj?func=openurl&genre=article&issn=13032968&date=2010&volume=9&issue=1&spage=98
2. W.S.S. Leite, Physiological demands in football, futsal and beach soccer: a brief review. Eur. J. Phys. Educ. Sport Sci. **2**, 1–10 (2016). https://doi.org/10.5281/ZENODO.205160
3. R. Sanitate, J. Harney, M. Schiro, D. Wollbrinck, M. Carrigg, C. Buell, Takraw: a global sport. Strategies **11**, 29–33 (1998)
4. M.N. Jawis, R. Singh, H.J. Singh, M.N. Yassin, Anthropometric and physiological profiles of sepak takraw players. Br. J. Sports Med. **39**, 825–829 (2005). https://doi.org/10.1136/bjsm.2004.016915
5. B. Ulmer, M. Fernandez, Predicting soccer match results in the English Premier League, 2013
6. C. Peace, E. Okechukwu, An improved prediction system for football a match result. IOSR J. Eng. **04**, 2250–3021 (2014)
7. C.P. Igiri, Support vector machine-based prediction system for a football match result. IOSR J. Comput. Eng. **17**, 21–26 (2015). https://doi.org/10.9790/0661-17332126
8. R.G. Martins, A.S. Martins, L.A. Neves, L.V. Lima, E.L. Flores, M.Z. do Nascimento, Exploring polynomial classifier to predict match results in football championships. Expert Syst. Appl. **83**, 79–93 (2017). https://doi.org/10.1016/J.ESWA.2017.04.040
9. A. Joseph, N.E. Fenton, M. Neil, Predicting football results using Bayesian nets and other machine learning techniques. Knowl.-Based Syst. **19**, 544–553 (2006). https://doi.org/10.1016/J.KNOSYS.2006.04.011

10. N. Razali, A. Mustapha, F.A. Yatim, R. Ab Aziz, Predicting football matches results using Bayesian networks for English Premier League (EPL). IOP Conf. Ser. Mater. Sci. Eng. **226**, 012099 (2017). https://doi.org/10.1088/1757-899X/226/1/012099

11. R.M. Musa, A.P.P. Abdul Majeed, M.A. Mohd Razman, M.A.H. Shaharudin, Match outcomes prediction of six top English Premier League clubs via machine learning technique, in *Communications in Computer and Information Science* (Springer Verlag, 2019), pp. 236–244. https://doi.org/10.1007/978-981-13-7780-8_20

12. A.E. Saricaoğlu, A. Aksoy, T. Kaya, Prediction of Turkish Super League match results using supervised machine learning techniques (2020). https://doi.org/10.1007/978-3-030-23756-1_34

13. G. Xiaohong, W. Yu, Analysis of basketball training model optimization based on artificial intelligence and computer aided model (2020). https://doi.org/10.1007/978-3-030-25128-4_257

14. M.S. Oughali, M. Bahloul, S.A. El Rahman, Analysis of NBA players and shot prediction using random forest and XGBoost models, in *2019 International Conference on Computer and Information Sciences, ICCIS 2019* (Institute of Electrical and Electronics Engineers Inc., 2019). https://doi.org/10.1109/ICCISci.2019.8716412

15. S. Valero, Predicting win-loss outcomes in MLB regular season games —a comparative study using data mining methods. Int. J. Comput. Sci. Sport **15** (2016). https://doi.org/10.1515/ijcss-2016-0007

16. B. Tolbert, T. Trafalis, Predicting Major League Baseball championship winners through data mining. Athens J. Sports **3**(4), 239 (2016). https://doi.org/10.30958/ajspo.3.4.1

17. N. Pathak, H. Wadhwa, Applications of modern classification techniques to predict the outcome of ODI cricket. Procedia Comput. Sci. 55–60 (2016). https://doi.org/10.1016/j.procs.2016.05.126

18. J. Kumar, R. Kumar, P. Kumar, Outcome prediction of ODI cricket matches using decision trees and MLP networks, in *ICSCCC 2018—1st International Conference on Secure Cyber Computing and Communications* (Institute of Electrical and Electronics Engineers Inc., 2019), pp. 343–347. https://doi.org/10.1109/ICSCCC.2018.8703301

19. M.M. Rahman, M.O.F. Shamim, S. Ismail, An analysis of Bangladesh One Day International cricket data: a machine learning approach, in *2018 International Conference on Innovations in Science, Engineering and Technology (ICISET)* (IEEE, 2018), pp. 190–194

20. R.M. Musa, M.R. Abdullah, A.B.H.M. Maliki, N.A. Kosni, M. Haque, The application of principal components analysis to recognize essential physical fitness components among youth development archers of Terengganu, Malaysia. Indian J. Sci. Technol. **9** (2016)

21. Z. Taha, M. Haque, R.M. Musa, M.R. Abdullah, A.B.H.M. Maliki, N. Alias, N.A. Kosni, Intelligent prediction of suitable physical characteristics toward archery performance using multivariate techniques. J. Glob. Pharma Technol. (2017)

22. M.R. Abdullah, A.B.H.M. Maliki, R.M. Musa, N.A. Kosni, H. Juahir, M. Haque, Multi-hierarchical pattern recognition of athlete's relative performance as a criterion for predicting potential athletes. J. Young Pharm. **8**, 463 (2016)

23. O. Maimon, L. Rokach, *Data Mining and Knowledge Discovery Handbook* (2005). https://doi.org/10.1007/b107408

24. R.M. Musa, M.R. Abdullah, A.B.H.M. Maliki, N.A. Kosni, S.M. Mat-Rasid, A. Adnan, H. Juahir, Supervised pattern recognition of archers' relative psychological coping skills as a component for a better archery performance. J. Fundam. Appl. Sci. **10**, 467–484 (2018)

25. R. Muazu Musa, A.P.P. Abdul Majeed, Z. Taha, M.R. Abdullah, A.B. Husin Musawi Maliki, N. Azura Kosni, The application of Artificial Neural Network and k-Nearest Neighbour classification models in the scouting of high-performance archers from a selected fitness and motor skill performance parameters. Sci. Sports (2019). https://doi.org/10.1016/j.scispo.2019.02.006

26. C. Wu, R.C. Gudivada, B.J. Aronow, A.G. Jegga, Computational drug repositioning through heterogeneous network clustering. BMC Syst. Biol. **7**, S6 (2013). https://doi.org/10.1186/1752-0509-7-S5-S6

27. V.D. Blondel, J. Guillaume, R. Lambiotte, E. Lefebvre, Fast unfolding of community hierarchies in large networks. J. Stat. Mech. Theory Exp. **2008** (2008). https://doi.org/10.1088/1742-5468/2008/10/P10008

28. A. Motevalli, S.A. Naghibi, H. Hashemi, R. Berndtsson, B. Pradhan, V. Gholami, Inverse method using boosted regression tree and k-nearest neighbor to quantify effects of point and non-point source nitrate pollution in groundwater. J. Clean. Prod. **228**, 1248–1263 (2019). https://doi.org/10.1016/j.jclepro.2019.04.293

29. M.A.M. Razman, G.A. Susto, A. Cenedese, A.P.P. Abdul Majeed, R.M. Musa, A.S. Abdul Ghani, F.A. Adnan, K.M. Ismail, Z. Taha, Y. Mukai, Hunger classification of *Lates calcarifer* by means of an automated feeder and image processing. Comput. Electron. Agric. **163** (2019). https://doi.org/10.1016/j.compag.2019.104883

30. K.J. Luken, R.P. Norris, L.A.F. Park, Preliminary results of using k-nearest neighbor regression to estimate the redshift of radio-selected data sets. Publ. Astron. Soc. Pacific. **131**, 108003 (2019). https://doi.org/10.1088/1538-3873/aaea17

31. F. Martínez, M.P. Frías, M.D. Pérez, A.J. Rivera, A methodology for applying k-nearest neighbor to time series forecasting. Artif. Intell. Rev. (2017). https://doi.org/10.1007/s10462-017-9593-z

32. T.M. Cover, P.E. Hart, Nearest neighbor pattern classification. IEEE Trans. Inf. Theory **13**, 21–27 (1967). https://doi.org/10.1109/TIT.1967.1053964

33. C. Cortes, V. Vapnik, Support-vector networks. Mach. Learn. **20**, 273–297 (1995)

34. W.S. McCulloch, W. Pitts, A logical calculus of the ideas immanent in nervous activity. Bull. Math. Biophys. **5**, 115–133 (1943). https://doi.org/10.1007/BF02478259

35. I.M. Yusri, A.P.P. Abdul Majeed, R. Mamat, M.F. Ghazali, O.I. Awad, W.H. Azmi, A review on the application of response surface method and artificial neural network in engine performance and exhaust emissions characteristics in alternative fuel. Renew. Sustain. Energy Rev. (2018). https://doi.org/10.1016/j.rser.2018.03.095

36. A. El-Sawy, A.P.P. Abdul Majeed, R.M. Musa, M.A. Mohd Razman, M.H.A. Hassan, A.A. Jaafar, The flexural strength prediction of porous Cu-Sn-Ti composites via artificial neural networks, in *Lecture Notes in Mechanical Engineering* (Pleiades Publishing, 2020), pp. 403–407. https://doi.org/10.1007/978-981-13-8323-6_34

37. M.A. Abdullah, M.A.R. Ibrahim, M.N.A.B. Shapiee, M.A. Mohd Razman, R.M. Musa, A.P.P. Abdul Majeed, The classification of skateboarding trick manoeuvres through the integration of IMU and machine learning (2020). https://doi.org/10.1007/978-981-13-9539-0_7

38. N.Q. Radzuan, M.H.A. Hassan, A.P.P. Abdul Majeed, R.M. Musa, M.A. Mohd Razman, K.A. Abu Kassim, Predicting serious injuries due to road traffic accidents in Malaysia by means of artificial neural network (2020). https://doi.org/10.1007/978-981-13-9539-0_8

39. L. Breiman, Random forests. Mach. Learn. **45**, 5–32 (2001)

40. M.R. Abdullah, R.M. Musa, A.B.H.M. Maliki, N.A. Kosni, P.K. Suppiah, Development of tablet application based notational analysis system and the establishment of its reliability in soccer. J. Phys. Educ. Sport **16**, 951–956 (2016). https://doi.org/10.7752/jpes.2016.03150

41. K. McGuigan, M. Hughes, D. Martin, Performance indicators in club level Gaelic football. Int. J. Perform. Anal. Sport **18**, 780–795 (2018). https://doi.org/10.1080/24748668.2018.1517291

Chapter 2
Key Performance Indicators in Elite Beach Soccer

2.1 Overview

Since the inception of the beach soccer sport and the subsequent recognition of the sport by FIFA, many researchers have sought to investigate various factors responsible for the successful performance in the sport [1]. These attempts are intended towards understanding the nature of the sport as well as the underlying variables that could aid performance and define its involvement at higher competitive tournaments. To this effect, a study of similarities and variations between football, beach soccer as well as futsal in world cup competitions was undertaken by previous researchers [2]. The researchers looked into the aggregate of the average number of goals scored, the circumstance under which the goals are scored with respect to the time intervals as well as the outcome of scoring the initial goal during a game and its subsequent effects towards the final results of match play. It was inferred from the findings of the study that the average number of goals scored in a game of futsal and football is bound to reduce whilst an increase in goals scored was witnessed in the game of beach soccer. Moreover, it was demonstrated that a considerable number of goals were scored in the last quarter of the time play. The chances of winning a game from scoring a first goal were observed to be about 70% in both futsal and football as opposed to the beach soccer which was found to be about 60%. It was concluded from the study that the differences of the performances from the study parameters were largely resulted from the tactical, technical, physical coupled with the physiological characteristics of each of the investigated sport and as such each sport differ with regard to the specific requirement of the game.

Scarfone et al. [3] carried out a study to investigate the match demands of beach soccer between professional and the amateur competitional championship. Matches of the Italian championship and a friendly match competed at the university level amounting to three were analysed by the authors. The physiological requirement of the game was assessed by means of heart monitor which was categorised as the

© The Author(s), under exclusive license to Springer Nature Singapore Pte Ltd. 2020
R. Muazu Musa et al., *Machine Learning in Team Sports*,
SpringerBriefs in Applied Sciences and Technology,
https://doi.org/10.1007/978-981-15-3219-1_2

percentage of the heart's maximum capacity, whilst the physical elements were evaluated through the application of time–motion analysis. Furthermore, the technical and tactical parameters were determined using a notational analysis system. It was demonstrated that the heart rate levels significantly differed in between the three matches investigated; nonetheless, no differences were detected between the two levels of expertise with regard to the time–motion analysis as well as the tactical and technical indicators. It was suggested that the distinction observed in the aforesaid parameters might be resulted from the tactical and the technical nature of the sport in which the players tend to rapidly switch from standing to sprinting particularly with regard to the heart rate readings, whilst the difficulty of playing in the sandy surface could be the possible reason for the lack of difference in the time–motion analysis. Finally, the researchers concluded that individualised-based evaluation technique could be considered as per the notational analysis system.

For any notational system to be effective, it is necessary to develop as well as define relevant performance indicators that could be used to evaluate individual or group performance. It is worth to note that evidence has demonstrated that coaches, as well as performance analysts, often utilise performance indicators to assess an individual or group performance with the sole aim of improving performance and provision of feedback that could enable the identification of strength or weakness [4]. Although many attempts have been made to investigate the physiology as well as physical demands of beach soccer games through motion analysis system, a little effort has been thus far devoted to establishing the key performance indicators that could be used to evaluate the team as well as individual performance in the elite beach soccer competitions. Therefore, this chapter aims to establish the essential performance indicators required for performance evaluation in elite beach soccer.

2.2 Development of Performance Indicators

A total number of 20 tactical as well as technical indicators were developed. The performance indicators that consist of shot back third, shot mid-third, shot front third, pass back third, pass mid-third, pass front third, shot inbox, shot outbox, chances created, interception, turnover, goals scored first period, goals scored second period, goals scored third period, goals scored extra time, tackling, fouls committed, complete save, incomplete save and passing error were developed to analyse the performance of the teams in the AFC beach soccer tournament. It is non-trivial to highlight that a number of these performance indicators were selected and considered based on their relevance to the beach soccer sport as documented in the previous studies [3, 5]. Moreover, it is worth noting that four experienced professional beach soccer coaches were responsible for validating the said performance indicators. The readers are encouraged to refer to the previous work for more detailed information about the operational definition of each performance indicator as well as the demarcation of the playing field [6].

2.3 The Application of Principal Component Analysis

In the present investigation, a principal component analysis (PCA) was used to identify the most essential performance indicators amongst the aforementioned indicators developed. The data gathered through the procedure explained in the previous chapter was used to carry out the analysis. In this study, a factor loading that is equal to or greater than 0.70 was considered as significant as such a variable that has lower than the threshold value set is eliminated [7–10]. It is important to note that prior to this analysis, the performance indicators were standardised to z-score transformation by scaling the mean and the standard deviation of all the indicators to 0.0 and 1.0, respectively. This process is considered apt in order to eliminate the effect of bias amongst the variables [11].

2.4 Results and Discussion

Figure 2.1 depicts the scree plot for the eigenvalues scores of the PCA. It could be observed from the figure that the PCA ascertained a total of seven components as key in explaining the entire dataset. These components are identified as essential due to the higher eigenvalues greater than 1 (>1) that are associated with the components. Based on this justification, therefore, these identified components were retained and subsequently used as inputs parameters for further analysis, i.e. varimax rotation as suggested by the precedent researchers [11, 12].

Table 2.1 demonstrates the PCA after applying the varimax rotation. It could be seen from the table that in each of the seven components, a number of key performance indicators are suggested. These suggested components are identified due to the fulfilment of the predetermined factor loading threshold previously set, i.e.

Fig. 2.1 A scree plot for the eigenvalue of the PCA

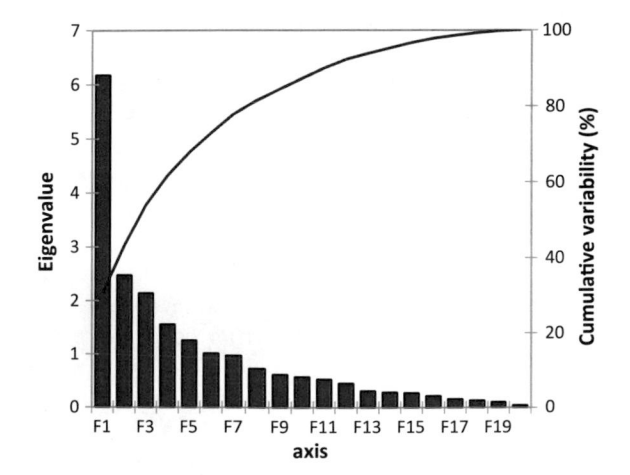

Table 2.1 PCA after varimax rotation

Indicators	C1	C2	C3	C4	C5	C6
SHOT BACK 1/3	−0.165	0.163	**0.785**	0.166	−0.091	0.143
SHOT MID 1/3	0.095	0.069	0.665	0.036	0.225	−0.252
SHOT FRONT 1/3	**0.720**	0.390	0.140	−0.096	0.080	0.215
PASS BACK 1/3	0.185	0.147	−0.028	**0.860**	0.019	0.037
PASS MID 1/3	0.481	0.373	0.105	0.302	−0.386	−0.320
PASS FRONT 1/3	0.540	0.616	0.041	0.197	0.064	0.252
SHOT IN BOX	0.249	**0.714**	0.449	0.018	0.217	0.189
SHOT OUT BOX	0.222	0.080	**0.845**	−0.208	0.020	−0.014
CHANCES CRTD	**0.706**	0.091	0.393	−0.118	0.069	0.230
INTERCEPTION	0.107	0.043	0.028	**−0.758**	0.042	−0.156
TURNOVER	−0.281	**−0.714**	−0.091	0.051	−0.069	−0.237
G.S. 1ST P	**0.759**	0.104	−0.010	0.124	−0.297	0.005
G.S. 2ND P	**0.759**	0.083	−0.024	0.194	−0.202	−0.012
G.S. 3RD P	0.348	0.155	0.048	0.201	−0.129	**0.772**
G.S. EXTR.T	0.046	0.207	0.311	0.060	**0.783**	0.013
TACKLING	0.532	−0.623	0.236	0.192	0.127	−0.130
FOULS COMM	−0.315	−0.698	−0.114	−0.153	−0.083	0.054
COMPLT S.KP	−0.030	**−0.806**	−0.083	−0.141	−0.003	0.155
UNCOMPLT S.KP	**−0.759**	−0.191	−0.022	0.005	−0.068	−0.267
PASSING ERR KP	−0.264	0.015	−0.486	−0.074	0.691	−0.232
Eigenvalue	6.175	2.472	2.135	1.552	1.256	1.011
Variability (%)	30.877	12.360	10.677	7.760	6.280	5.056
Cumulative (%)	30.877	43.237	53.914	61.674	67.954	73.010

greater or equal to the 0.70 that are shown in bold. It could be observed that a total number of 14 performance indicators out of the 20 initially developed are identified by all the seven components as key for the evaluation of performance in elite beach soccer.

The performance indicators identified as key are shot at front third, chances created, goals scored at the first period, goals scored at the second period, goals scored at third period, goals scored at extra time, shot at back third, pass at back third, interception, turnover, complete save by the keeper, uncomplete save by the keeper, shot inside a box as well as shot outside box. These set of performance indicators identified when broken down could be interpreted as technical and tactical strategies that could ensure wining as well as errors that could be attributed to a losing performance during a match play.

Performance indicators, such as goals scored at all the four periods of match play (first, second, third and extra time), chances created, passes at back and front third, shot at back third, inbox as well as outbox identified in the present investigation, could be attributed to technical skills. It has been documented that the nature of the soccer game is multi-faceted and as such a successful delivery of performance in the game is reliant upon the technical ability of a player that could aid in developing an awareness of the game, decision-making whilst striking the ball, obtaining the ball with both foot, passing the ball with different parts of the body, manoeuvering away from the opposition, dribbling the ball at certain speed as well as controlling the ball from the air [6, 8, 13–15]. This set of skills could be a potential discriminator between elite and novice players. Therefore, it is crucial to understand why the technical skill is considered as essential for performance evaluation in elite beach soccer competitions.

In the present investigation, interception and complete save by the keeper are identified as essential for evaluating performance in elite beach soccer as demonstrated in Table 2.1. These sets of performance indicators could be interpreted as a tactical ability. A typical beach soccer game is identified with a steady bouncing coupled with the unexpected transfer of the ball into many directions. These factors necessitate that a player is required to possess tactical awareness which involves the ability to know positional role in the pitch, quick response to intercepting a pass as well as a good vision by a goalkeeper to anticipate shots from any direction of the field [15–17]. Conversely, the inability of a goalkeeper to save shots from the opposition players as well as turnover or losing possession of the ball by the outfield players could lead to an ineffective performance in this sport. It is, therefore, not surprising that uncompleted saves, as well as turnover, are depicted as important indicators that could be employed to assess the performance of beach soccer in the present investigation. This finding is congruent with the findings of the previous researchers who reported that lack of tactical awareness by both goalkeepers as well as outfield players could result in losing a game during beach soccer tournaments [6, 18, 19].

2.5 Summary

The present investigation has established the key performance indicators that could be potentially employed to analyse the team as well as players' performance in the elite beach soccer tournament. A set of performance indicators, namely shot at front third, chances created, goals scored at first period, goals scored at second period, goals scored at third period, goals scored at extra time, shot at back third, pass at back third, interception, turnover, complete save by the keeper, uncomplete save by the keeper, shot inside a box as well as shot outside box, is demonstrated to be essential for evaluating performance. It has been shown that technical–tactical as well as error-related indicators should be considered when designing a notational analysis system in elite beach soccer tournaments.

References

1. J. Castellano, D. Casamichana, Heart rate and motion analysis by GPS in beach soccer. J. Sports Sci. Med. **9**, 98–103 (2010). http://www.doaj.org/doaj?func=openurl&genre=article& issn=13032968&date=2010&volume=9&issue=1&spage=98
2. W.S.S. Leite, Physiological demands in football, futsal and beach soccer: a brief review. Eur. J. Phys. Educ. Sport Sci. **2**, 1–10 (2016). https://doi.org/10.5281/ZENODO.205160
3. R. Scarfone, C. Minganti, A. Ammendolia, L. Capranica, A. Tessitore, Match analysis and heart rate of beach soccer players during amateur competition. Grad. J. Sport Exerc. Phys. Educ. Res. **3**, 1–12 (2015)
4. R.M. Musa, M.R. Abdullah, A.B.H.M. Maliki, N.A. Kosni, S.M. Mat-Rashid, A. Adnan, N. Alias, V. Eswaramoorthi, The effectiveness of tablet-based application as a medium for feedback in performance analysis during a competitive match in elite soccer. Movement Health Exerc. **6**, 57–65 (2017)
5. A. Bravo-Sánchez, J. Abián-Vicén, P. Abián, Analysis of the physical and technical differences between 7-a-side and 8-a-side game modalities in official under 12 soccer matches. Int. J. Perform. Anal. Sport **17**, 545–554 (2017). https://doi.org/10.1080/24748668.2017.1366760
6. R. Muazu Musa, A.P.P. Abdul Majeed, M.R. Abdullah, A.F.A. Nasir, M.H. Arif Hassan, M.A. Mohd Razman, Technical and tactical performance indicators discriminating winning and losing team in elite Asian beach soccer tournament. PLoS One **14**, e0219138 (2019). https://doi.org/10.1371/journal.pone.0219138
7. R.M. Musa, M.R. Abdullah, A.B.H.M. Maliki, N.A. Kosni, M. Haque, The application of principal components analysis to recognize essential physical fitness components among youth development archers of Terengganu, Malaysia. Indian J. Sci. Technol. **9** (2016)
8. M.R. Abdullah, R.M. Musa, N.A. Kosni, A.B.H.M. Maliki, M.S.A.A. Karim, M. Haque, Similarities and distinction pattern recognition of physical fitness related performance between amateur soccer and field hockey players. Int. J. Life Sci. Pharma Res. **6**, L35–L46 (2016)
9. M.R. Razali, N. Alias, A. Maliki, R.M. Musa, L.A. Kosni, H. Juahir, Unsupervised pattern recognition of physical fitness related performance parameters among Terengganu youth female field hockey players. Int. J. Adv. Sci. Eng. Inf. Technol. **7**, 100–105 (2017)
10. Z. Taha, M. Haque, R.M. Musa, M.R. Abdullah, A.B.H.M. Maliki, N. Alias, N.A. Kosni, Intelligent prediction of suitable physical characteristics toward archery performance using multivariate techniques. J. Glob. Pharma Technol. (2017)
11. S. Shrestha, F. Kazama, Assessment of surface water quality using multivariate statistical techniques: a case study of the Fuji river basin, Japan. Environ. Model. Softw. **22**, 464–475 (2007). https://doi.org/10.1016/j.envsoft.2006.02.001
12. M.R. Abdullah, A.B.H.M. Maliki, R.M. Musa, N.A. Kosni, H. Juahir, M. Haque, Multihierarchical pattern recognition of athlete's relative performance as a criterion for predicting potential athletes. J. Young Pharm. **8** (2016)
13. M. Hughes, T. Caudrelier, N. James, A. Redwood-Brown, I. Donnelly, A. Kirkbride, C. Duschesne, Moneyball and soccer—an analysis of the key performance indicators of elite male soccer players by position. J. Human Sport Exerc. **7**, 402–412 (2012). https://doi.org/10.4100/jhse.2012.72.06
14. R. Scarfone, A. Tessitore, C. Minganti, A. Ferragina, A. Ammendolia, Match demands of beach soccer between amateur and professional competition, in *International Conference of Sport Science* (2010), pp. 8–10
15. W. Leite, D. Barreira, Are the teams sports soccer, futsal and beach soccer similar? Int. J. Sports Sci. Coach. **4**, 75–84 (2014). https://doi.org/10.5923/s.sports.201401.11
16. K. McGuigan, M. Hughes, D. Martin, Performance indicators in club level Gaelic football. Int. J. Perform. Anal. Sport **18**, 780–795 (2018). https://doi.org/10.1080/24748668.2018.1517291
17. R. Scarfone, A. Tessitore, C. Minganti, A. Ferragina, L. Capranica, A. Ammendolia, Match demands of beach soccer: a case study, in *Book of Abstracts of 14th Annual Congress of the European College of Sport Science* (2009), pp. 24–29

18. 2014, F., *Technical Report and Statistics CM 2014* (2010)
19. W.S.S. Leite, Beach soccer: analysis of the goals scored and its relation to the game physiology. Int. J. Phys. Educ. Fit. Sports **5**, 12–17 (2016). https://doi.org/10.26524/1613

Chapter 3
Technical and Tactical Performance Indicators Determining Successful and Unsuccessful Team in Elite Beach Soccer

3.1 Overview

Evidence has demonstrated that performance analysis is often utilised as a diagnostic tool for the assessment of team or players' performances [1–3]. The employment of performance analysis to quantify as well as discriminate successful and unsuccessful performances has been extensively studied in various sports. A few of such studies are the identification of defensive strategies in the 2010 FIFA World Cup, the establishment of performance parameters in club-level Gaelic football, development of notational analysis system in elite soccer and its corresponding reliability, respectively [4–6]. A different study also investigated the essential performance indicators in elite male soccer with respect to a different position of players [2]. A similar study looked into the performance indicators that could distinguish between the winning and losing teams in the sport of beach volleyball [7].

In beach soccer sport, a number of researchers in different studies have investigated the physiological as well as the physical demand of play in the sport through the integration of motion analysis coupled with heart rate evaluations [8–11]. Injury occurrences in the sport have also been given undue attention [12, 13]. A few of the aforementioned studies have considered integrating some performance analysis elements associated with the demand of the game. A more related study that investigated some performance analysis variables has reported that in approximately 52.8% of the inversion play used in beach soccer, at least a number of two players are involved. Similarly, about 42.6% of the attacking strategy is executed by a single pass. It was revealed in the study that beach soccer can be considered as intermittent in nature which required a steady supply of both aerobic and anaerobic energies [14].

There are considerably limited data on the performance indicators that could discriminate between successful and unsuccessful teams in beach soccer sport, and thus the aim of the present study is providing relevant data on the performance indicators that could potentially determine successful as well as losing performances.

© The Author(s), under exclusive license to Springer Nature Singapore Pte Ltd. 2020   21
R. Muazu Musa et al., *Machine Learning in Team Sports*,
SpringerBriefs in Applied Sciences and Technology,
https://doi.org/10.1007/978-981-15-3219-1_3

3.2   Clustering

In the present study, a set of essential performance indicators that were established via principal component analysis in the previous chapter (Chap. 2), namely shot at front third, chances created, goals scored at the first period, goals scored at the second period, goals scored at third period, goals scored at extra time, shot at back third, pass at back third, interception, turnover, complete save by the keeper, uncomplete save by the keeper, shot inside a box as well as shot outside box, were used as the dataset. The Louvain clustering algorithm is employed to group the performances of the teams. Two classes were formed, i.e. successful team (SUC-T) and unsuccessful team (UNS-T) based on their respective performances on the evaluated performance indicators. Moreover, the Mann–Whitney U test was applied to determine the statistical differences between the SUC-T and UNS-T with regard to the performance indicators assessed in the study.

3.3   Classification

In this chapter, artificial neural network (ANN) with a single hidden layer is used in order to classify successful and unsuccessful teams. The features used are obtained from the previous analysis. In this study, a number of hyperparameters are optimised by means of the grid search, namely the number of hidden neurons, activation functions as well as the learning algorithm. The number of neurons within the hidden layer was varied between 10 and 50 with an interval of 10 neurons. The type of activation functions was varied between the rectified linear unit (ReLu), tangent sigmoid (tanh) and logistic, whilst the learning algorithm was varied between adaptive, constant and inverse scaling (invscaling). The data was split into training, validation and testing with a ratio of 70:30:30 from a total of 45 observations. The fivefold cross-validation technique was used on the training data, in order to identify the best combination of the hyperparameters that could yield a reasonably well classification accuracy. Python's scikit-learn library through the Spyder platform was used in this study.

3.4   Results and Discussion

Figure 3.1 displays the grouping defined by the Louvain clustering algorithm. It could be observed from the figure that a somewhat clear demarcation is created between the successful and the unsuccessful teams. This separation of the groups provided evidence that the playing strategies between the teams differ with respect to the execution of the performance indicators evaluated in the present study.

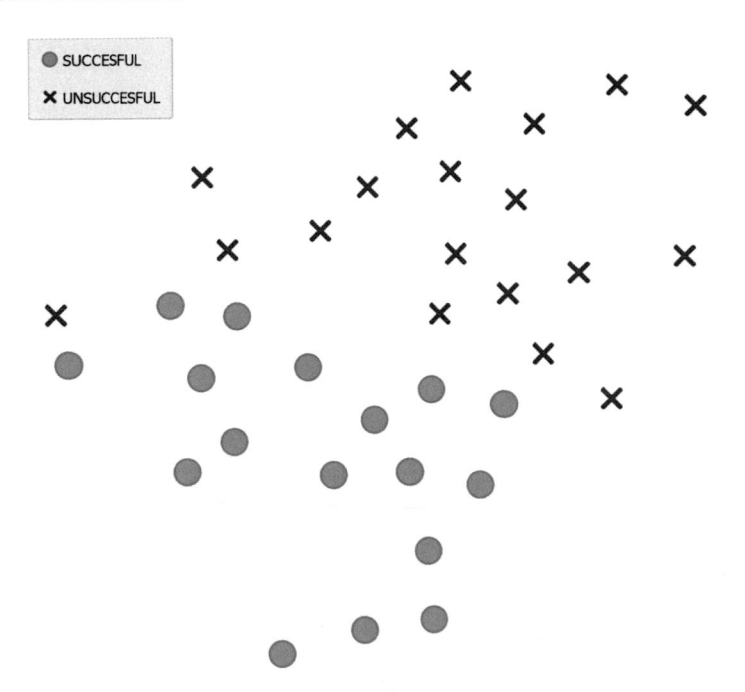

Fig. 3.1 Performance groups defined by Louvain clustering algorithm

Table 3.1 demonstrates the Mann–Whitney U test conducted with regard to the performance indicators assessed. It could be seen that a total of seven performance indicators out of the 14 initially evaluated were statistically different between the two teams $p < 0.05$. These sets of performance indicators, viz. pass back third, shot back third, interception, turnover, goals scored in the second period, goals scored in the third period, as well as complete saves by the keeper, are shown to be the major performance indicators that could potentially distinguish between the SUC-T

Table 3.1 Inferential statistics of the Mann–Whitney U test for the indicators evaluated between the teams

Performance indicators	SUC-T		UNS-T		
	Mean	Std. dev.	Mean	Std. dev.	P-value (MWU)
Pass back third	59.75	13.074	28.52	8.856	0.001
Shot back third	8.25	5.159	5.92	3.402	0.021
Interception	8.95	4.524	12.92	4.751	0.021
Turnover	13.65	7.322	18.32	8.586	0.032
Goals scored second period	1.85	1.725	0.96	1.136	0.049
Goals scored third period	1.85	1.461	0.96	0.935	0.033
Complete save by keeper	9.05	4.571	13.76	6.009	0.005

and the UNS-T. It is worthy to note that these performance indicators constitute the combination of both defensives and attacking play.

Figure 3.2 provides the visual presentation of the average performance difference between the two teams via boxplots. It could be noticed that the mean performances of the successful teams are higher in all the performance indicators except for the turnover, complete save as well as interceptions. It is, therefore, apparent from the boxplots' visualisations that unsuccessful teams are engrossed with more attacks from the opposition teams and thus were compelled to defend more as compared to the successful teams.

From the grid search analysis, the optimal combination of the hyperparameters is 20, tanh and invscaling, for the number of hidden neurons, activation function and learning algorithm, respectively. It was established from the features selected and the optimised ANN model that the classification accuracy (CA) of the model for the training, validation and testing is 100%, 100% and 85%, respectively. The confusion matrix for the train, validation and test dataset is depicted in Figs. 3.3, 3.4 and 3.5, respectively. It is worth noting that the model does not overfit as the CA for validation data is also 100%, although the CA for the test data is 85%; nonetheless, it could be seen from the confusion matrix (Fig. 3.5) that only one misclassification transpired, suggesting the integrity of the model developed is not compromised.

3.5   Summary

The present chapter investigated the influence of technical and tactical performance indicators in determining the match outcome, i.e. successful and unsuccessful teams in elite beach soccer. It was established from the current investigation that some technical as well as tactical performance indicators, namely pass back third, shot back third, interception, turnover, goals scored in the second period, goals scored in the third period, as well as complete saves by the keeper, could potentially differentiate between winning and losing performances in elite beach soccer. It is worth to note that the aforesaid performance indicators involve the assortments of both defensive and offensive strategies in the sport. Thus, these indicators identified could be considered when designing a performance analysis system for the evaluation of players' performances in the sport. Moreover, the application of a non-conventional technique particularly artificial neural networks in classifying the performance groups is demonstrated to be valuable towards the provision of a reasonably good prediction model of both the successful and unsuccessful teams with respect to the performance indicators identified in the current investigation.

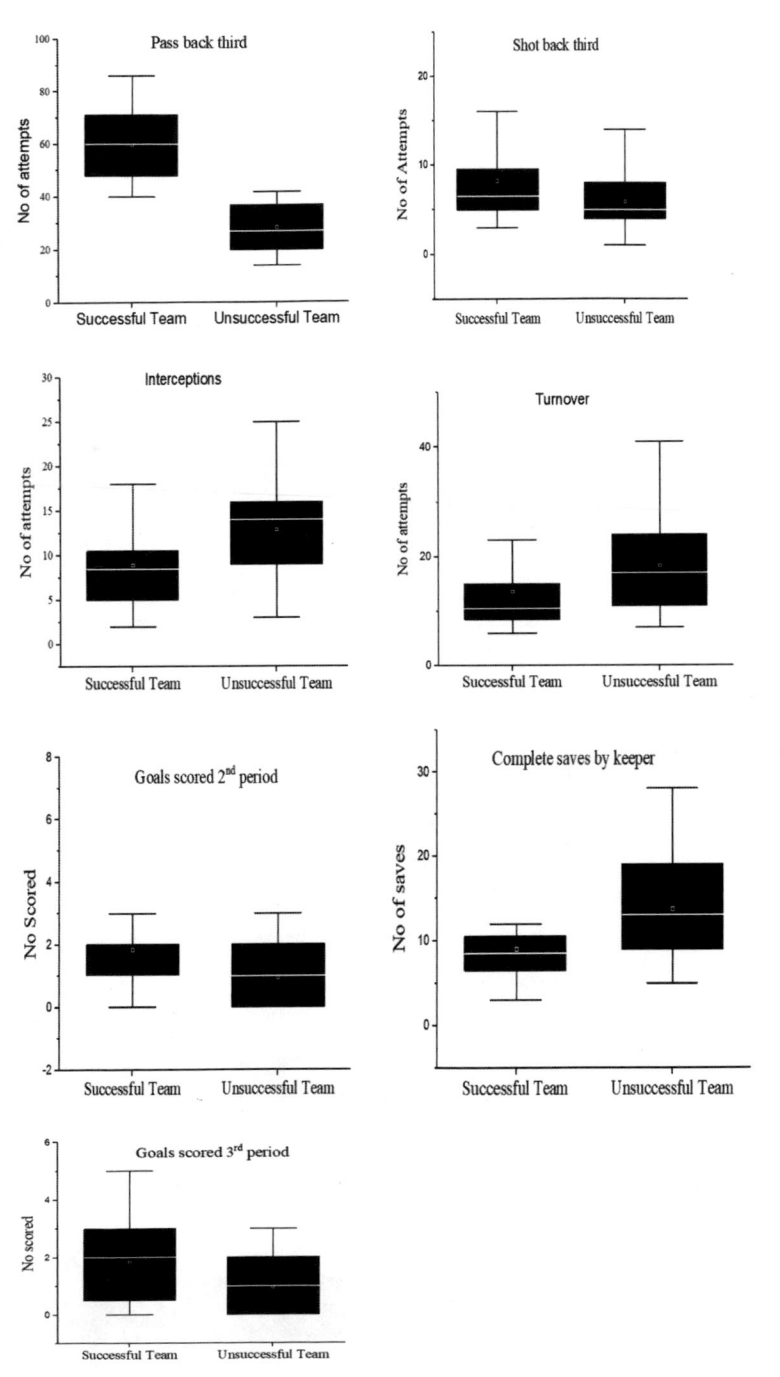

Fig. 3.2 Performance differences between the successful and unsuccessful teams from the seven indicators

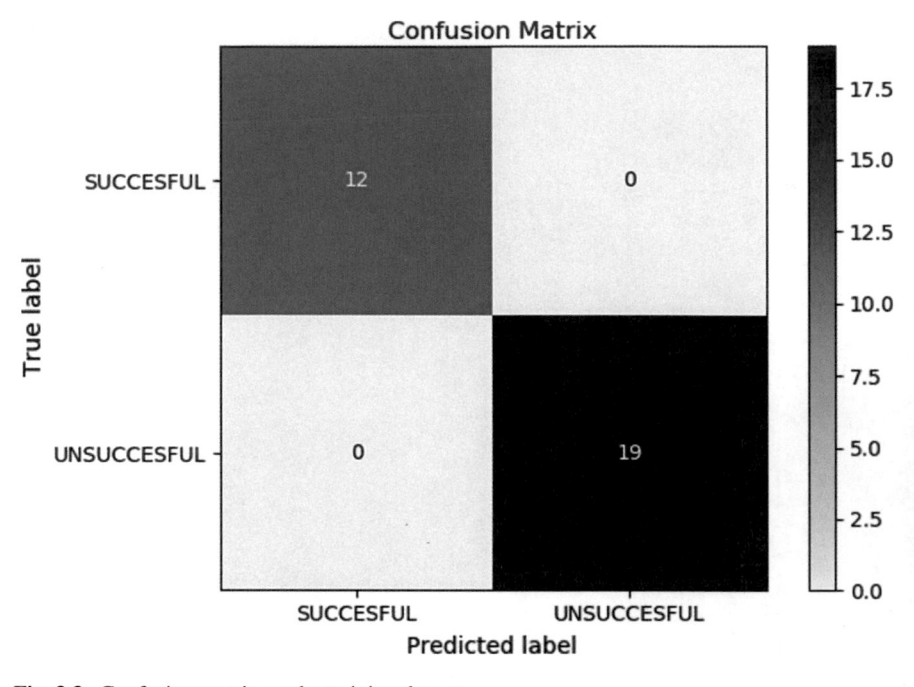

Fig. 3.3 Confusion matrix on the training dataset

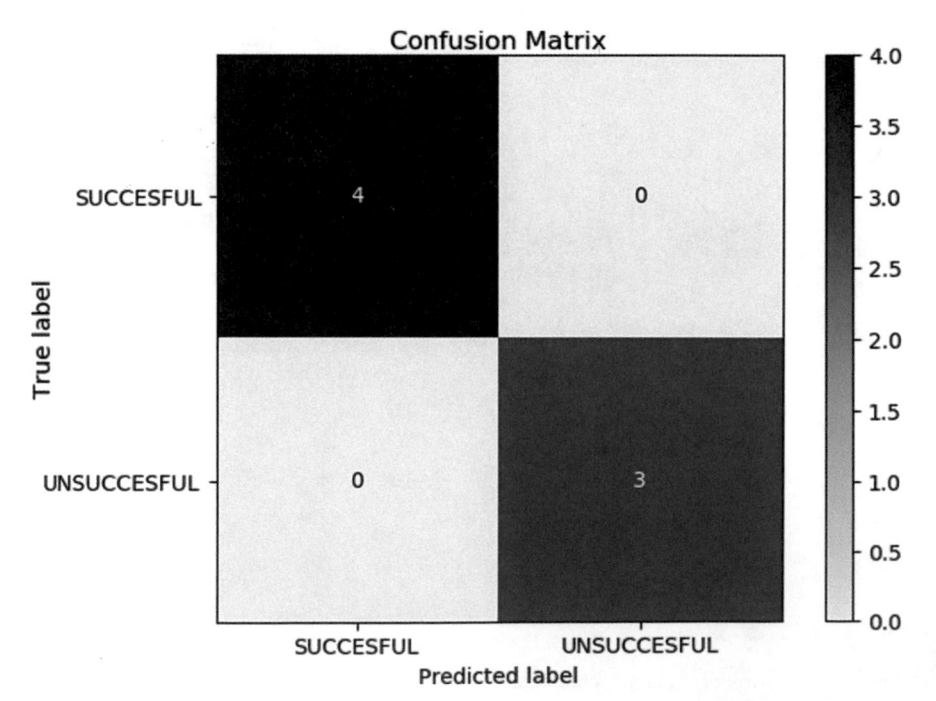

Fig. 3.4 Confusion matrix on the validation dataset

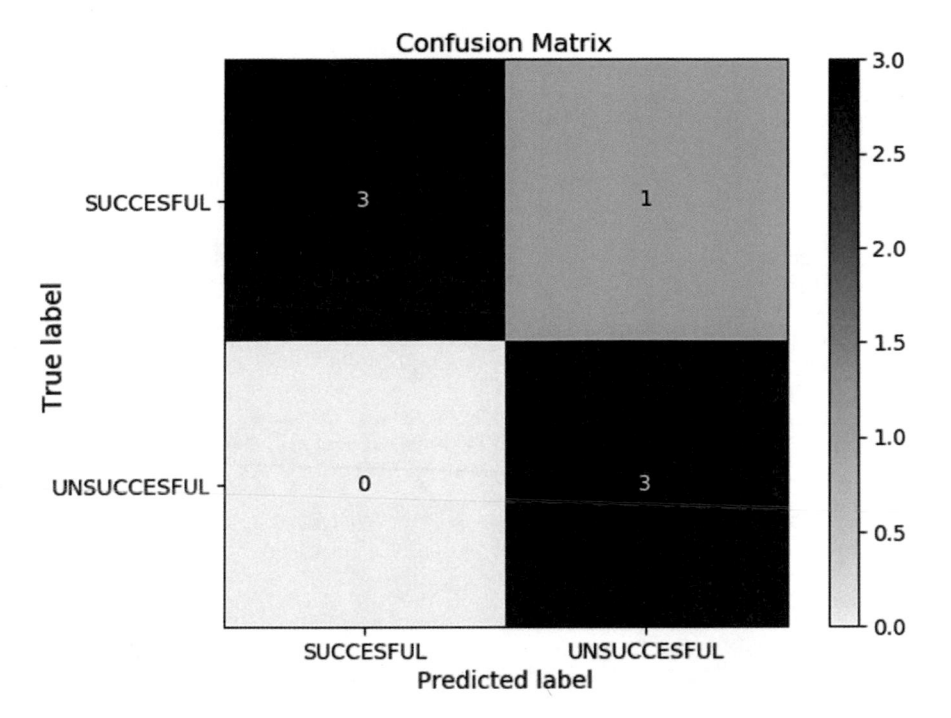

Fig. 3.5 Confusion matrix on the test dataset

References

1. A. Arnason, S.B. Sigurdsson, A. Gudmundsson, I. Holme, L. Engebretsen, R. Bahr, Physical fitness, injuries, and team performance in soccer. Med. Sci. Sports Exerc. **36**, 278–285 (2004). https://doi.org/10.1249/01.MSS.0000113478.92945.CA
2. M. Hughes, T. Caudrelier, N. James, A. Redwood-Brown, I. Donnelly, A. Kirkbride, C. Duschesne, Moneyball and soccer—an analysis of the key performance indicators of elite male soccer players by position. J. Hum. Sport Exerc. **7**, 402–412 (2012). https://doi.org/10.4100/jhse.2012.72.06
3. M.A. Gomez, C. Lago-Peñas, J. Viaño, I. González-Garcia, Effects of game location, team quality and final outcome on game-related statistics in professional handball close games. Kinesiol. Int. J. Fundam. Appl. Kinesiol. **46**, 249–257 (2014)
4. C. Casal, M. Andujar, J. Losada, T. Ardá, R. Maneiro, Identification of defensive performance factors in the 2010 FIFA world cup South Africa. Sports **4**, 54 (2016). https://doi.org/10.3390/sports4040054
5. K. McGuigan, M. Hughes, D. Martin, Performance indicators in club level Gaelic football. Int. J. Perform. Anal. Sport. **18**, 780–795 (2018). https://doi.org/10.1080/24748668.2018.1517291
6. M.R. Abdullah, R.M. Musa, A.B.H.M. Maliki, N.A. Kosni, P.K. Suppiah, Development of tablet application based notational analysis system and the establishment of its reliability in soccer. J. Phys. Educ. Sport. **16**, 951–956 (2016). https://doi.org/10.7752/jpes.2016.03150
7. A.I.A. Medeiros, R. Marcelino, I.M. Mesquita, J.M. Palao, Performance differences between winning and losing under-19, under-21 and senior teams in men's beach volleyball. Int. J. Perform. Anal. Sport. **17**, 96–108 (2017). https://doi.org/10.1080/24748668.2017.1304029

8. R. Scarfone, C. Minganti, A. Ammendolia, L. Capranica, A. Tessitore, Match analysis and heart rate of beach soccer players during amateur competition. Grad. J. Sport. Exerc. Phys. Educ. Res. **3**, 1–12 (2015)
9. W.S.S. Leite, Beach soccer: analysis of the goals scored and its relation to the game physiology. Int. J. Phys. Educ. Fit. Sport. **5**, 12–17 (2016). https://doi.org/10.26524/1613
10. R. Scarfone, A. Tessitore, C. Minganti, A. Ferrragina, L. Capranica, A. Ammendolia, Match demands of beach soccer: a case study, in *Book of Abstracts of 14th Annual Congress of the European College of Sport Science* (2009), pp. 24–29
11. S. Scarfone, A. Tessitore, C. Minganti, A. Ferragana, A. Ammedolia, Match demands of beach soccer between amateur and professional competition, in *International Conference of Sport Science* (2010), pp. 8–10
12. S. Al-Shaqsi, A. Al-Kashmiri, A. Al-Risi, S. Al-Mawali, Sports injuries and illnesses during the second Asian beach games. Br. J. Sports Med. **46**, 780–787 (2012). https://doi.org/10.1136/bjsports-2011-090852
13. T. Shimakawa, Y. Shimakawa, Y. Kawasoe, K. Yoshimura, Y. Chinen, K. Eimon, W. Chibana, S. Shirota, K. Kadekawa, R. Bahr, T. Uezato, H. Ikeda, Beach soccer injuries during the Japanese national championships. Orthop. J. Sport. Med. **4**, 2325967115625636 (2016). https://doi.org/10.1177/2325967115625636
14. R. Scarfone, A. Ammendolia, Match analysis of an elite beach soccer team. J. Sport. Med. Phys. Fit. **57**, 953–959 (2017). https://doi.org/10.23736/S0022-4707.16.06580-4

Chapter 4
Identifying Talent in Sepak Takraw via Anthropometry Indexes

4.1 Overview

Anthropometry characteristics could be defined as the attributes or features that explain the physical as well as the anatomical disposition of an individual [1]. The anthropometric index of an individual consists of the composition, size as well as the shape of the individual. The study of anthropometry characteristics could be dated back to early 1920s, although the attribution of the tests measurements could vary; however, anthropometry attributions such as height, weight, per cent body fat as well as strength have been extensively used by numerous researchers to examine the variations of such attributes with regard to a diverse sporting activity [2–4]. A study has also demonstrated that some anthropometry features such as shapes, sizes as well as the composition of an individual could be used to forecast the strength of a person which consequently resulted in the determination of motor performances [4]. Moreover, it is to note that a positive significant association between strength and anthropometry of an individual in a number of sports has been previously documented in a previous study [5]. The study has further demonstrated that stronger athletes could be attributed with a certain body type that comprises particular body size, composition as well as indexes that somewhat differentiate them in the performance of the motor activity.

Individuals are naturally characterised on the basis of their anatomical as well as their body shape or otherwise referred to as body somatotype. It is important to highlight that in most cases, bodily characteristics are inherited from the family. For instance, the likelihood of having taller children is higher when the parents are tall and vice versa. Physical characteristics coupled with body composition have been reported to be essential in the classification as well as discrimination of athletic performances as previously noted. Conversely, it is non-trivial to mention that body

© The Author(s), under exclusive license to Springer Nature Singapore Pte Ltd. 2020
R. Muazu Musa et al., *Machine Learning in Team Sports*,
SpringerBriefs in Applied Sciences and Technology,
https://doi.org/10.1007/978-981-15-3219-1_4

type of an individual may not necessarily determine or predict performance in some sports. Nonetheless, some success has been reported in the identification as well as classification of talents via the employment of anthropometry features in a number of studies in different sports [4, 6–8]. The sport of sepak takraw could be considered as multi-faceted in nature due to the numerous demands of the sports in performances since it combines both the ability to sustain balance and strength to counter any attack from the opposition. The successful delivery of performance in the sport might, therefore, be connected to the anatomical structure coupled with the body composition of the players. It is against this background that the present study is undertaken to establish the relationship between the anthropometry indexes of sepak takraw players and performance delivery with the aim of identifying talent via the physical attributions of the players.

Anthropometry evaluations: The standard protocols for the assessments of anthropometry were adhered to in the present investigation. Anthropometric indexes, specifically the standing, sitting height, leg length, waist circumference, thigh circumference, calf circumference, and four-site skinfold measurements were carried out. The standing height was measured via a stadiometer, whilst the sitting height was determined from the maximum height of the head to the seated buttocks. The weight was measured using an electronic digital scale. The calf, hip, thigh, waist, as well as the leg lengths were assessed through the utilisation of brittle measuring tape. A skinfold calliper was used to measure the triceps, biceps, suprailiac and subscapular of the players. The performance ability of the players was determined from five basic skills of sepak takraw game, namely service, spiking, blocking, break ball as well as passing. The percentage of the successful performance of the skills was used for statistical analysis.

4.2   Clustering

The Louvain clustering algorithm is used to cluster the performance of the players with respect to the anthropometric indexes evaluated in the present investigation. A total of three clusters were formed which are coined as high-performance players (HPP), medium performance players and low-performance players (LPP). It is essential to note that the assignation of the class membership was made based on the anthropometric indexes as well as the corresponding performance ability of the players. In other words, the higher- and the medium-performance groups were coined due to their high-performance ability in conjunction with relatively bigger body physique. The lower-performance group is categorised with lower performance ability and a comparatively smaller body size.

4.3 Classification

In the present investigation, support vector machine (SVM) is employed to classify the classes of the sepak takraw players to either LPP, MPP or HPP, respectively. The dataset consists of 74 observations, in which a ratio of 70:15:15 was divided for training, validation and testing, respectively. The fivefold cross-validation technique was utilised during the training phase. The significant anthropometric indexes identified are employed to develop the model. It is worth noting that the SVM is further fine-tuned through optimising the hyperparameters by means of grid search. In this study, different kernels are investigated, namely linear, radial basis function (RBF) as well as a cubic polynomial function. Moreover, the range of the slack variable (soft margin) penalty term, C was varied from 0.1 to 1000 with a multiplication interval of 10, whilst the range of gamma is varied from 0.001 to 1 with the same multiplication factor. It is worth noting that the gamma term is applicable for both the polynomial and RBF-based SVM models only. The performance metrics evaluated during the model development as well as evaluation is the classification accuracy. Python's scikit-learn library through the Spyder platform was used in this study.

4.4 Results and Discussion

Figure 4.1 depicts the group memberships assigned by the Louvain clustering algorithm. It could be observed from the figure that the clustering algorithm is able to provide clear segregation of the performance classes of the players. This segregation of the groups demonstrated that the anthropometric indexes, as well as the performance ability of the sepak takraw players, differed from each other.

Figure 4.2 exhibits the graphical illustrations of the average performance difference between the three defined classes through boxplots. It could be noted that both the HPP and MPP are attributed to considerably higher body sizes as well as the performance ability of the sepak takraw as opposed to the LPP group. It could, therefore, be inferred from the boxplot visualisations that successful delivery of performance in sepak takraw might be attributed to bigger body sizes.

It was established from the grid search analysis on the training dataset that the hyperparameters of C, gamma and kernel of 0.1, 0.0001 and polynomial attained the highest classification accuracy of 88.2%. Figure 4.3 illustrates the grid search topology for the polynomial kernel. The optimised SVM model is then evaluated with both the validation and the test dataset, and classification accuracy of 100% and 92% was attained, respectively. It is apparent from the results obtained from the validation and test dataset that overfitting of the model did not transpire, suggesting the robustness of the developed model. Figures 4.4, 4.5 and 4.6 illustrate the confusion matrix for the train, validation and test datasets. The LPP, MPP and HPP are denoted as 0, 1 and 2, respectively. It could be observed from Fig. 4.6 that only a single misclassification transpired on the test dataset, i.e. MPP misclassified as HPP.

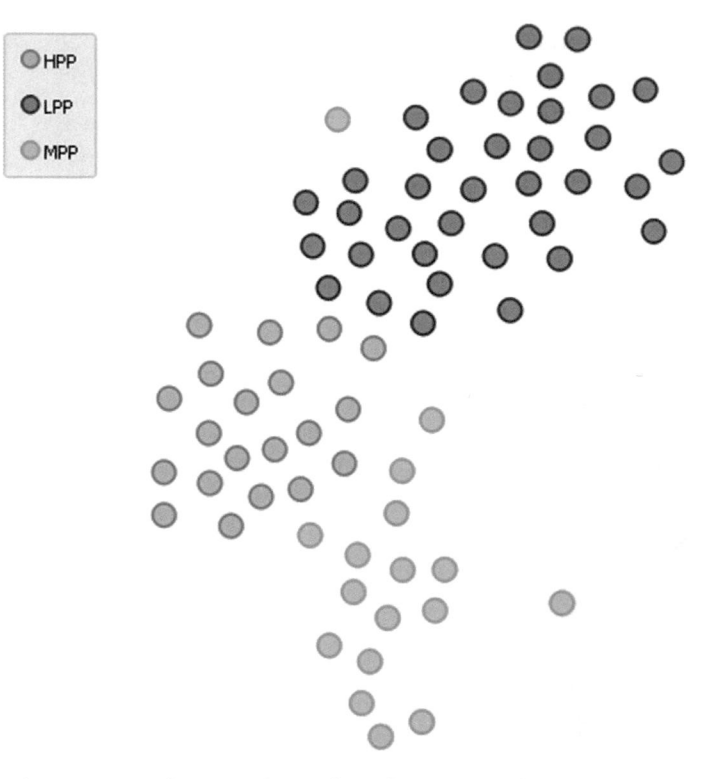

Fig. 4.1 Anthropometry performance clusters formed

4.5   Summary

The overall findings from the present chapter demonstrated that successful delivery of performance in sepak takraw sport could be attributed to the physical characteristics of the players. It is evident from the study findings that anthropometric indexes, specifically standing height, sitting height, leg length, waist circumference, thigh circumference, calf circumference, and four-site skinfold measurements (triceps, biceps, suprailiac and subscapular) investigated do affect performance in sepak takraw, and hence, players with relatively bigger body physique are shown to perform better compared to their counterparts. Conversely, the application of SVM is demonstrated to be effective in classifying the performance groups of the players with a noteworthy classification accuracy of 100% and 92% for both validation and test datasets, respectively. This finding has further highlighted that talent in the sport of sepak takraw could be identified by means of anatomical disposition of a player.

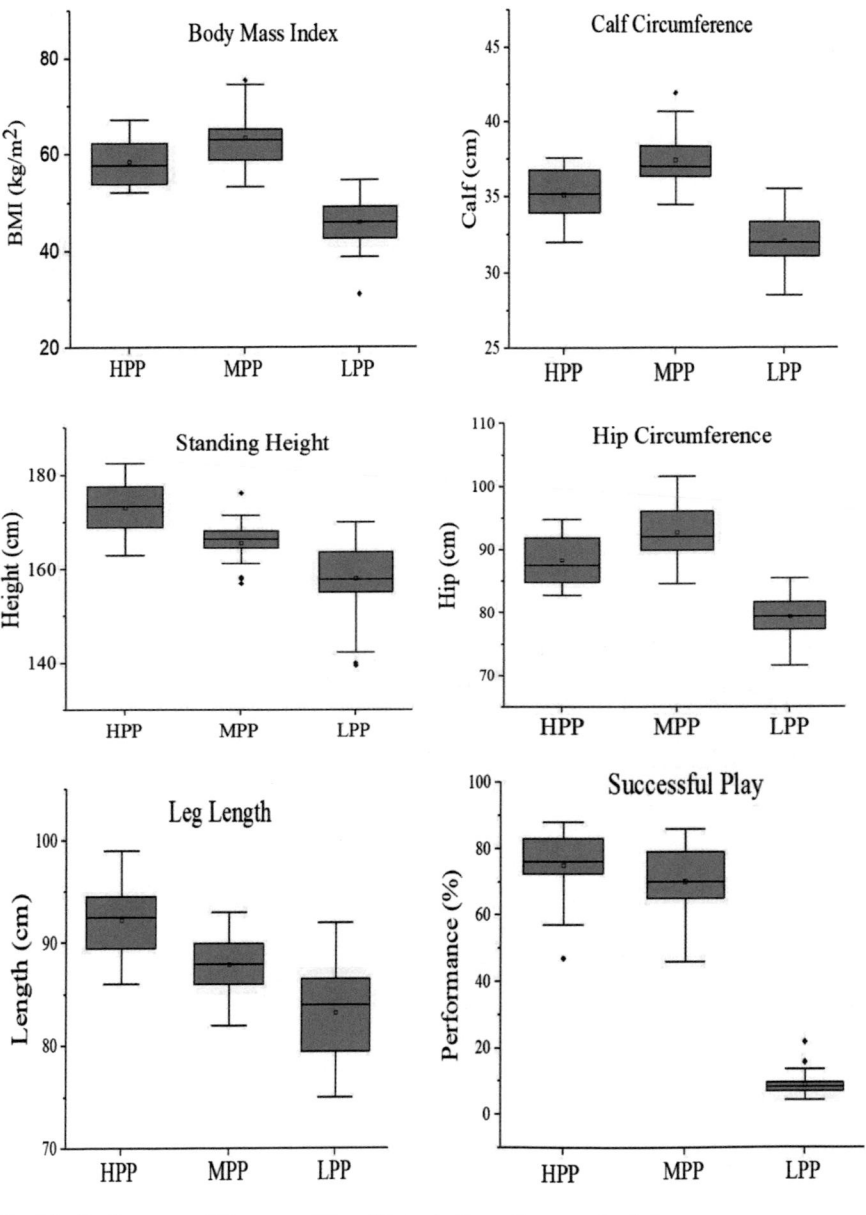

Fig. 4.2 Performance differences of the athletes in the anthropometry indexes

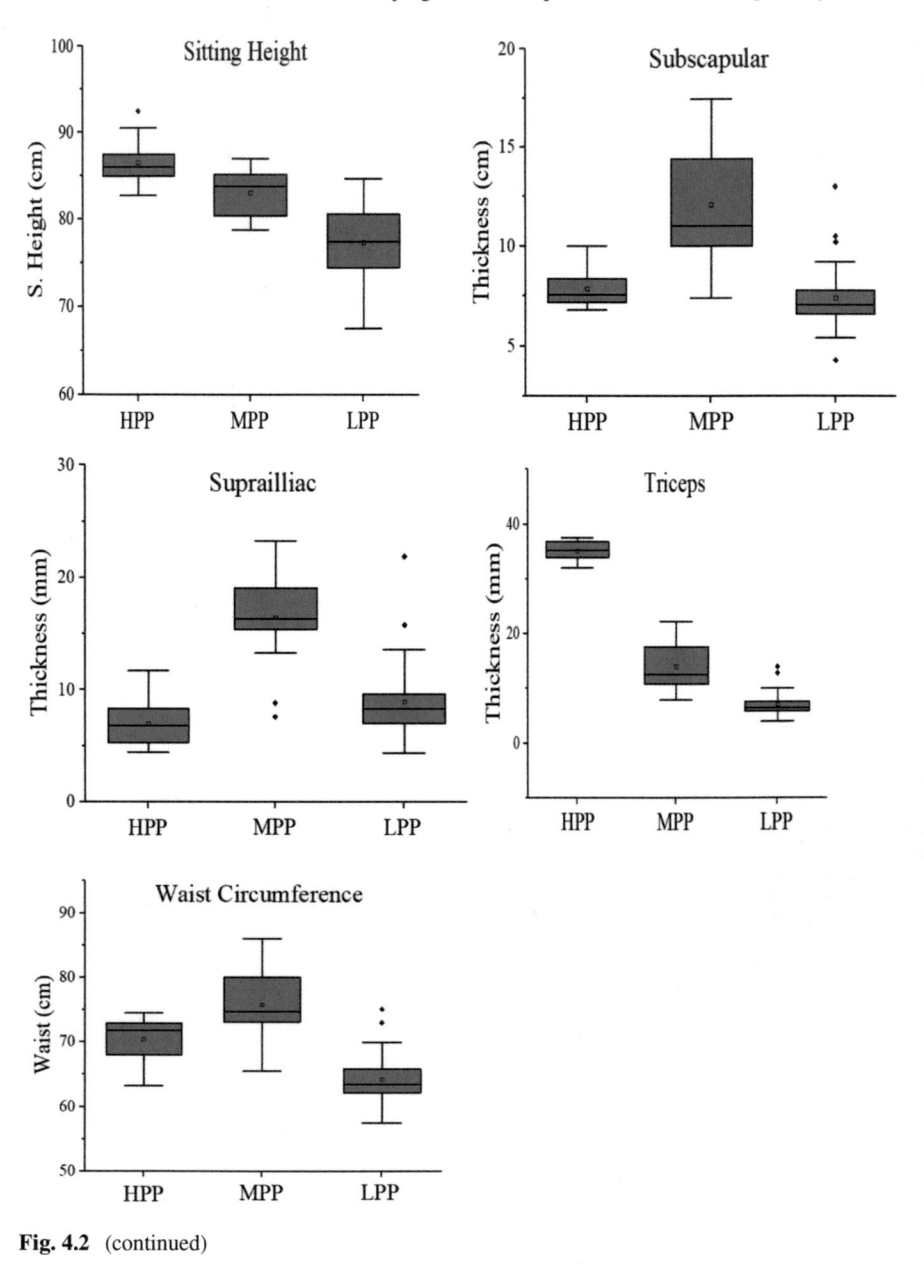

Fig. 4.2 (continued)

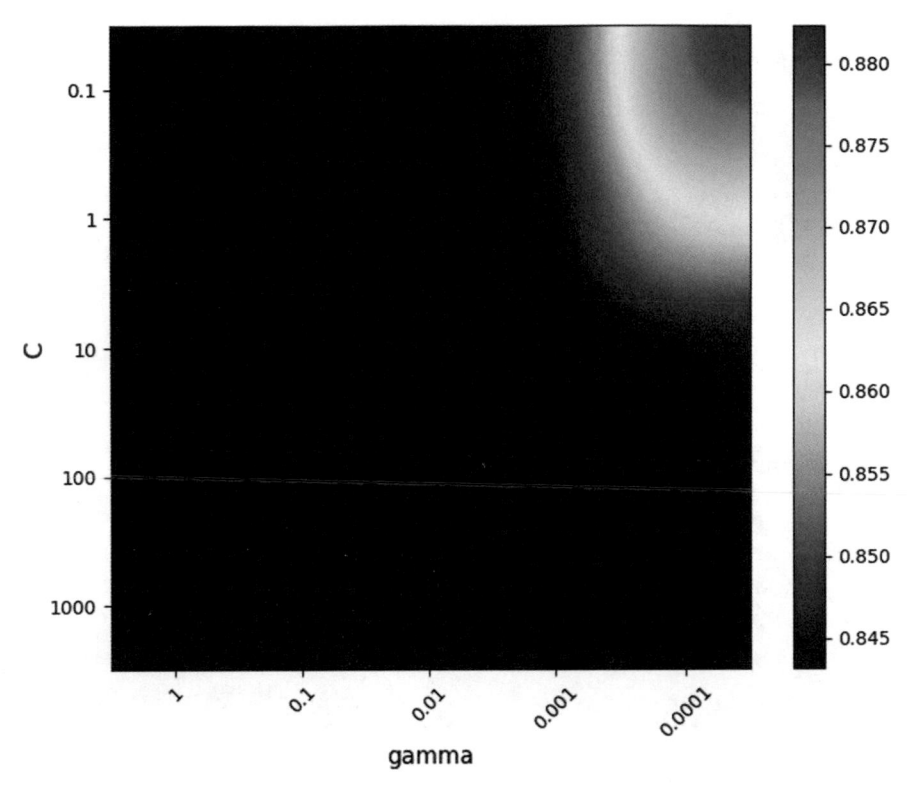

Fig. 4.3 Grid search topology for the polynomial kernel

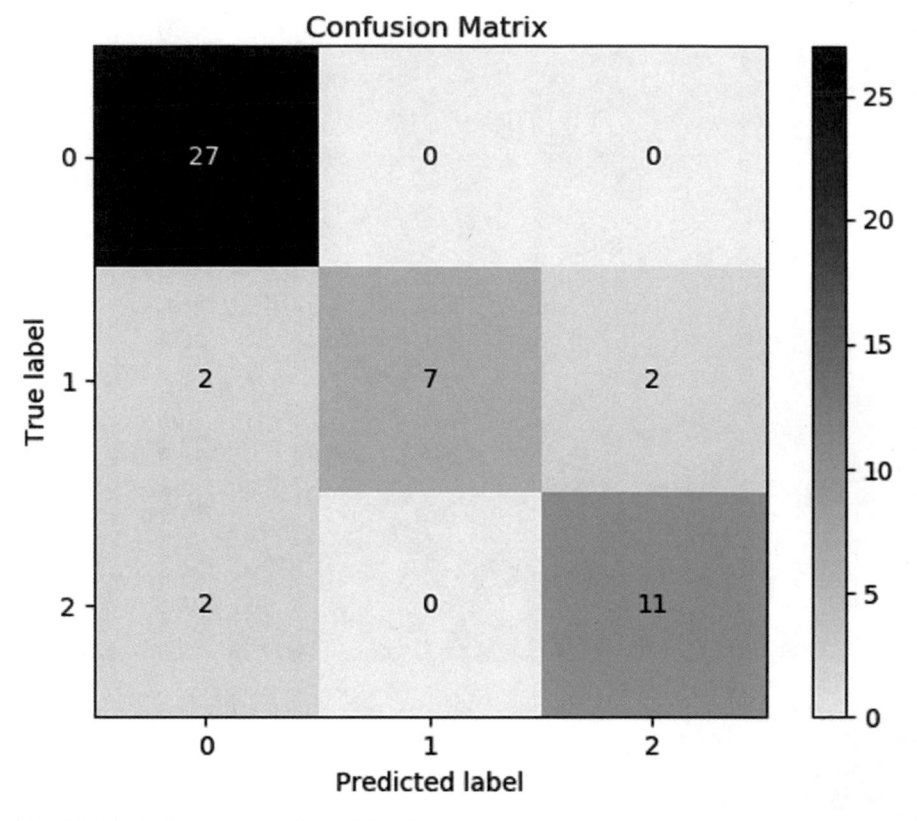

Fig. 4.4 Confusion matrix on the training dataset

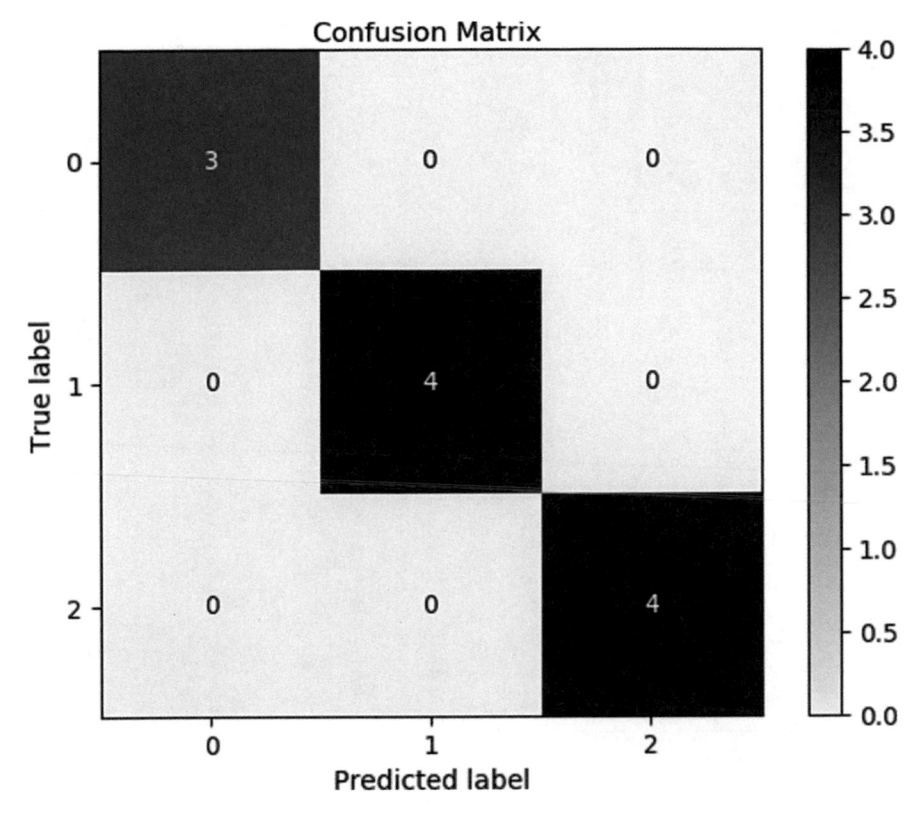

Fig. 4.5 Confusion matrix on the validation dataset

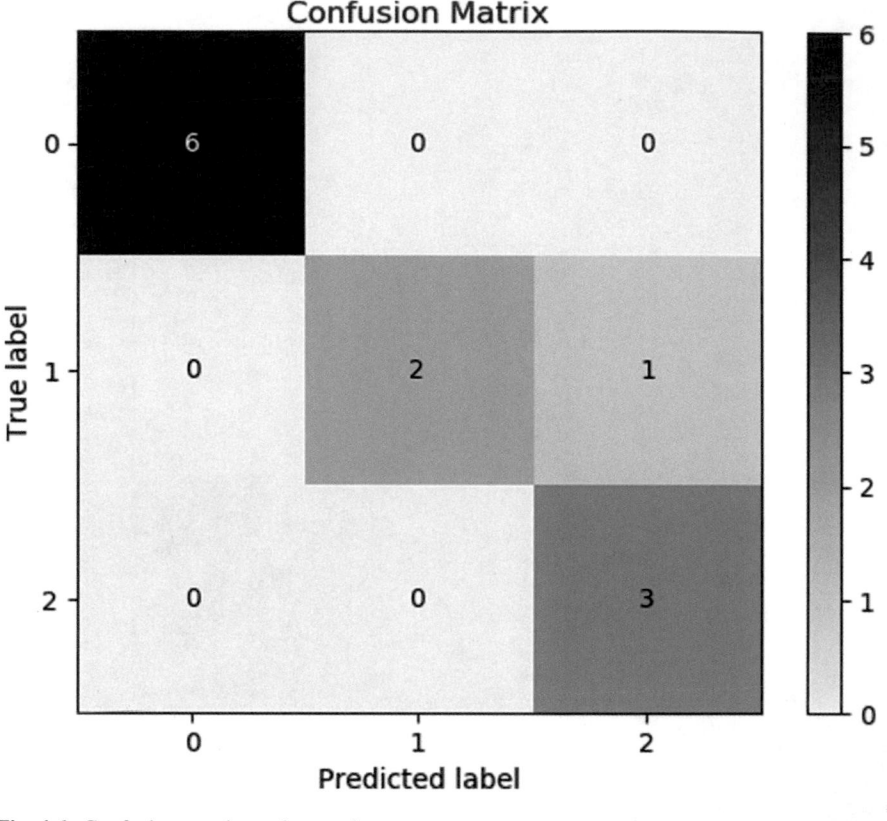

Fig. 4.6 Confusion matrix on the test dataset

References

1. R.M. Musa, Z. Taha, P.A. Majeed, Abdullah, M.R.: Anthropometry correlation towards archery performance, in *SpringerBriefs in Applied Sciences and Technology* (2019), pp. 29–35. https://doi.org/10.1007/978-981-13-2592-2_4
2. D.B. Pyne, G.M. Duthie, P.U. Saunders, C.A. Petersen, M.R. Portus, Anthropometric and strength correlates of fast bowling speed in junior and senior cricketers. J. Strength Cond. Res. **20**(3), 620 (2006)
3. T.J. Gabbett, Physiological and anthropometric characteristics of elite women rugby league players. J. Strength Cond. Res. **21**, 875–881 (2007)
4. S.M. Ostojic, S. Mazic, N. Dikic, Profiling in basketball: physical and physiological characteristics of elite players. J. Strength Cond. Res. **20**, 740–744 (2006)
5. T.E. Ball, B.H. Massey, J.E. Misner, B.C. Mckeown, T.G. Lohman, The relative contribution of strength and physique to running and jumping performance of boys 7-11. J. Sports Med. Phys. Fitness **32**, 364–371 (1992)
6. S. Amri, A.F. Ujang, M.R.W.N. Wazir, A.N. Ismail, Anthropometric correlates of motor performance among Malaysian university athletes. Mov. Heal. Exerc. 1 (2012)

7. R.M. Musa, M.R. Abdullah, A.B.H.M. Maliki, N.A. Kosni, M. Haque, The application of principal components analysis to recognize essential physical fitness components among youth development archers of Terengganu, Malaysia. Indian J. Sci. Technol. **9** (2016)
8. H.-B. Kim, S.-H. Kim, W.-Y. So, The relative importance of performance factors in Korean archery. J. Strength Cond. Res. **29**, 1211–1219 (2015). https://doi.org/10.1519/JSC. 0000000000000687

Chapter 5
Physical Fitness Parameters in the Identification of High-Potential Sepak Takraw Players

5.1 Overview

Physical fitness is considered as one of the prerequisites for successful performance in any sporting activity. Physical fitness aids athletes to recruit several muscles as well as structures that are involved in the performance of human movements. A significant number of researchers have stressed the importance of physical fitness in assisting athletes to perform well and consequently excel amongst their peers [1–4]. The sport of sepak takraw in nature has constituted the combination of several other sports such as soccer, volleyball, gymnastics as well as basketball to mention a few; these combined characteristics of the sport necessitate that a diverse range of physical fitness attributes are required for a sepak takraw player to deliver the best performance. Thus, some skill-related components of physical fitness such as agility, balance, coordination as well as explosive power could be essential to the successful performance in the sport of sepak takraw. Leg is the predominant body part used in performing the skills required in the sport, therefore, the ability of a sepak takraw player to change direction quickly and in a precise manner (agility), the capacity to maintain equilibrium whilst changing direction (balance), the capacity to direct the ball to a specific location (coordination) as well as the ability to hit the ball with a significant amount of force (power) could to a larger extent assists the players to cope with the demand of the game and as such the higher a player possesses the aforesaid elements the better the delivery of performance.

Since the introduction of sepak takraw game in the 10th Asian Games that held in Beijing in 1990 as well as the presentation of the sport as a demonstration in the 1998 Commonwealth Games in Kuala Lumpur, the sport of sepak takraw has gained enormous popularity and the game has grown and spread faster in both Asia and over 20 countries in different parts of the world [5]. Following the popularity and the admiration of the sport shown by many countries, a number of researchers began looking at the nature of the sport in an attempt to offer insights into the characteristics as well as the demands inherent in participating in the sport. The physiological and

R. Muazu Musa et al., *Machine Learning in Team Sports*,
SpringerBriefs in Applied Sciences and Technology,
https://doi.org/10.1007/978-981-15-3219-1_5

anthropometrics profile of the Malaysian sepak takraw players were investigated by the preceding researchers [5]. It was inferred from the study that the average weight, height and the cardiopulmonary capabilities of the players are within the range of the Malaysian population criterion; nonetheless, the investigated variables were comparatively found to be lower than some players in some other countries. Moreover, a study of physical and physiological demands of the game with respect to elite Singaporean players' profiles was carried out by other researchers [6]. The temporal nature of matches, namely duration of matches, rally, recovery and set play coupled with the physiological responses such as heart rate and blood lactate during matches, was investigated. The overall findings of the study revealed that the players' fitness characteristics are congruent with their physiological responses as well as the temporal characteristics that are involved in the match play.

A number of studies ranging from anthropometric characteristics with respect to match playing ability, comparative analysis of juggling skills, hamstring injury occurrences and its relationship with physical characteristics of collegiate sepak takraw players as well as head impact power of sepak takraw ball on sepak takraw players were also investigated [7–10]. Although many attempts have been made in different aspects of the sport by numerous researchers, there appears to be limited or no attention thus far made to look into the association of physical fitness parameters specifically dynamic balance, speed, peak power, explosive power and static balance with respect to the sepak takraw playing ability in a view to identify potential youth sepak takraw players. As such, the present investigation embarked on identifying potential sepak takraw players with regard to the aforesaid fitness parameters.

Specific fitness parameters measurements: A vertical jump test was used to assess the peak power velocity of the players. The jumping ability of the players was ascertained and subsequently converted into a peak power velocity which has been reported as one of the essential motor skills required in lower-body-based sports [11, 12]. A standing broad jump test was used to measure the explosive power of the players. A double take-off procedure in all the tests was used, and the players were permitted a maximum of three trials whereby the average scores were computed as per suggested by the previous researchers [4, 13]. A 20 m speed was used to determine the maximal speed of the players, whilst the star excursion balance test (SEBT) and the stork satance balance test was carried out to ascertain the dynamic and static balancing ability of the players, respectively. It is worth noting that all the aforementioned evaluations were conducted in accordance with the standard procedures of physical fitness testing described by the previous researchers [14, 15].

5.2 Clustering

In the present investigation, the hierarchical agglomerative cluster analysis (HACA) was employed to group the players based on their performance in the measured physical fitness parameters, namely dynamic balance, speed, peak power, explosive power and static balance. The overall performances of the players in the measured fitness parameters coupled with the pre-determined sepak takraw performance ability of the players were used by the HACA to construct as well as assign class membership to each player.

5.3 Classification

In this study, k-nearest neighbour (k-NN) classifier was utilised to ascertain the classes of the players previously clustered. The dataset that consists of 74 observations was split into the 70:15:15 ratio for training, testing and validation purpose. During the training phase, the fivefold cross-validation technique was employed. It is worth noting that one of the hyperparameters for k-NN, particularly the number of k, varied from 1 to 25 to identify its optimum value. This k is tuned by evaluating the classification accuracy of the model. In this study, the distance metric applied is the Euclidean distance. Python's scikit-learn library through the Spyder platform was used in this study.

5.4 Results and Discussion

Figure 5.1 demonstrates the performance group provided by the HACA. It could be seen from the figure that a separation of the performance classes is displayed through

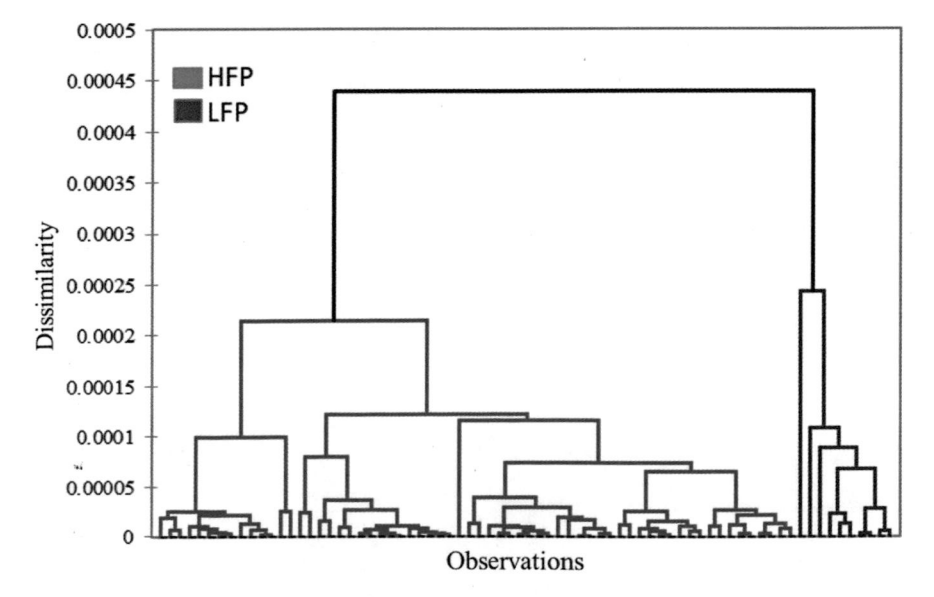

Fig. 5.1 Performance classes of the players defined by HACA

the dendrogram. The HACA was able to separate the players into two different classes based on the evaluated physical fitness parameters. The grouping of the players enables the assignation of the performance class for each player, i.e. the players who recorded higher performances in the assessed parameters are coined as high-fit players (HFP), whereas the players who obtained lower performances are named as low-fit players (LFP). It was established from the grouping of the players that a total number of 63 players are grouped as HFP, whilst 11 players are found to be within the class of LFP.

The graphical presentation of the mean differences between the two defined classes is shown in Fig. 5.2. It could be observed from the boxplots in the figure that the mean performance of the HFP is greater in all the evaluated physical fitness parameters as well as the sepak takraw playing ability assessed in the study.

The best attributes identified are then used to train the model. Figure 5.3 depicts the search for the optimum k-value from the training accuracy against the k plot. It could be observed from the plot that the training accuracy plateaued at $98 \pm 0.079\%$ from k is equal to three (3) onwards before it suffers considerably from 15 onwards by considering the fivefold cross-validation technique. Therefore, from the plot, it could be established that $k = 3$ yields the best classification model.

The trained model is then tested on both the validation (11 observations) and testing (12 observations) sets to evaluate the efficacy and the robustness of the developed model in classifying unseen data. It is noteworthy to highlight that the model is able to yield an excellent classification accuracy of 100% for both validation and

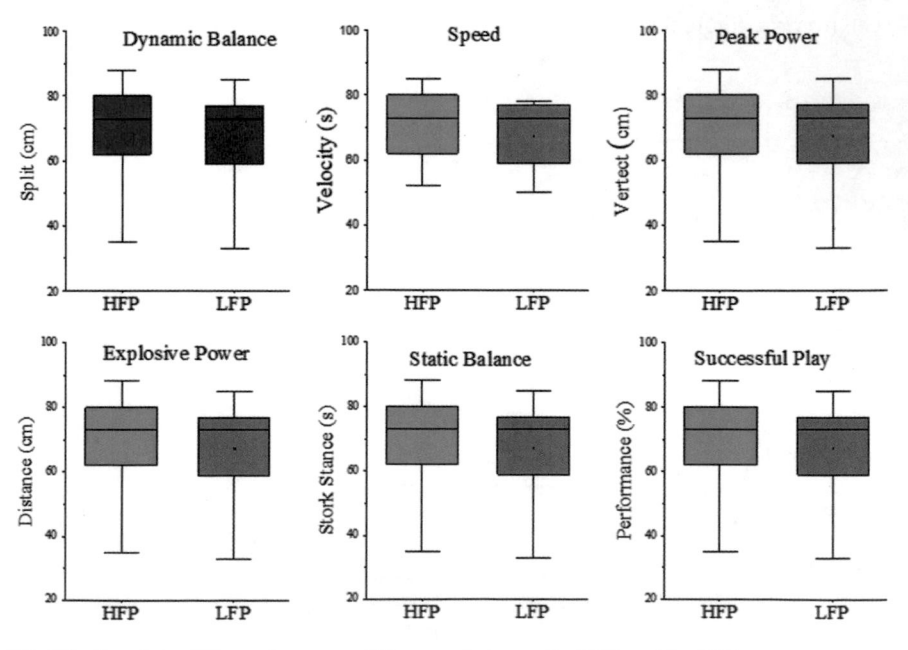

Fig. 5.2 Boxplots of the performance differences between the HFP and the LFP

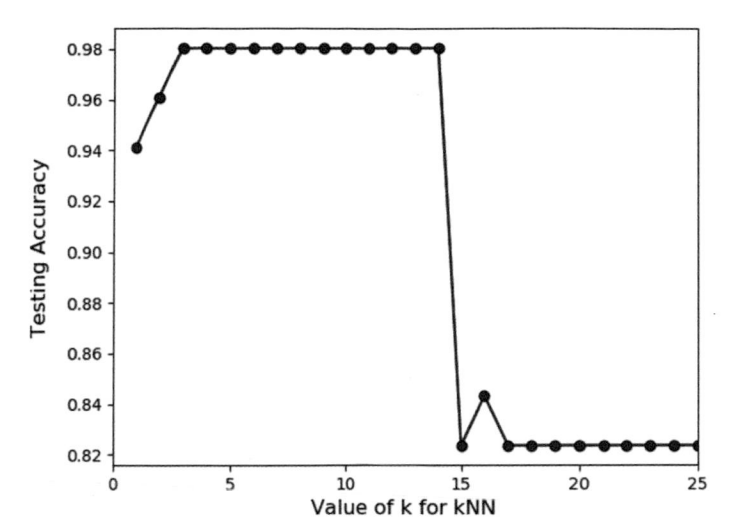

Fig. 5.3   Training accuracy against k curve

test data, respectively, suggesting the model developed does not exhibit undesirable overfitting behaviour on the data. Figures 5.4 and 5.5 illustrate the confusion matrix for both the validation and test data, respectively.

5.5   Summary

The requirement of physical fitness in a sporting activity has been stressed in many kinds of literature as noted earlier. Physical fitness is an integral element that plays a major role in many sports particularly aerobic-based towards the execution of sport-specific movement as well as aid in meeting up with the physical demands inherent with aerobic sports. The sport of sepak takraw is aerobic based in nature that might place a greater demand for physical prowess to the players. The results from the current investigation have demonstrated that successful performance in the sport of sepak takraw could be influenced by certain physical fitness attributes. Specific fitness parameters, namely peak power velocity, explosive power, speed, dynamic as well as static balance, are shown to be essential in the determination of successful performance of the sport. It has been revealed from the present investigation that high-performance sepak takraw players could be identified through their performance in the aforesaid fitness parameters. Moreover, the study findings highlighted that via the utilisation of the k-NN learning algorithm, an excellent classification accuracy could be attained. The excellent classification accuracy obtained in the study has underscored the importance of the selected fitness parameters in the study such that high-performance players could be classified and recognised accurately.

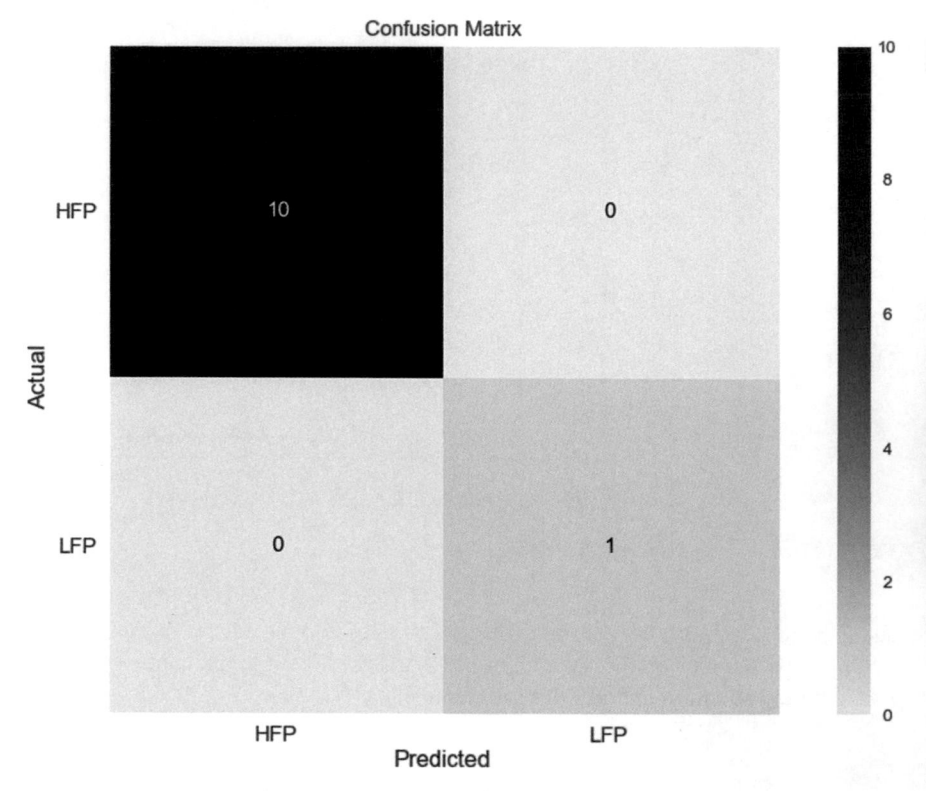

Fig. 5.4 Confusion matrix for the validation set

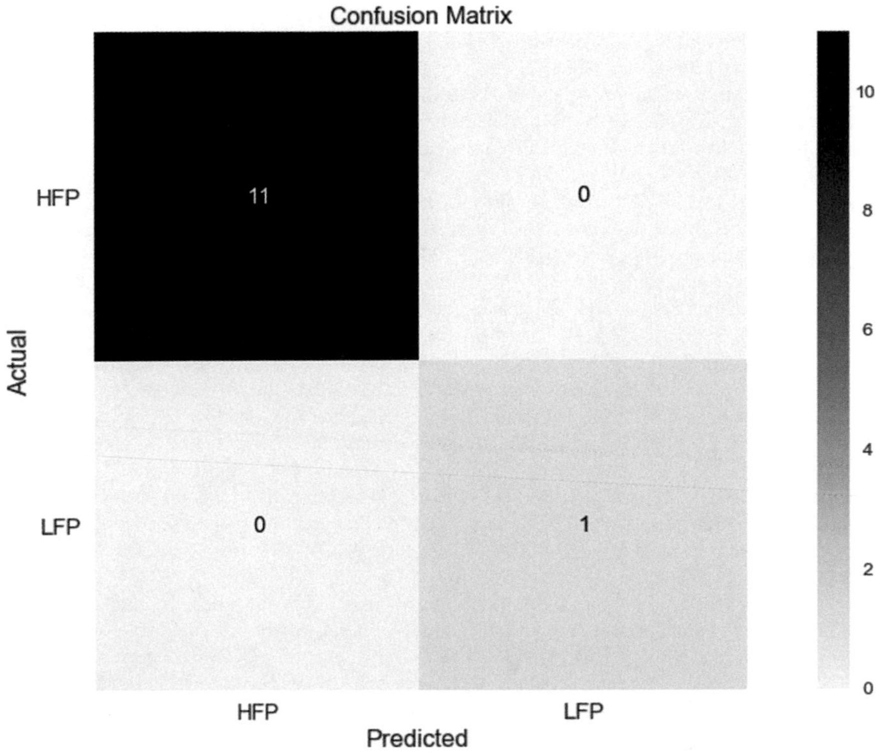

Confusion Matrix

	HFP	LFP
HFP	11	0
LFP	0	1

Actual / Predicted

Fig. 5.5 Confusion matrix for the testing set

References

1. R. Muazu Musa, A.P.P.A. Majeed, Z. Taha, M.R. Abdullah, A.B. Husin Musawi Maliki, N. Azura Kosni, The application of artificial neural network and k-nearest neighbour classification models in the scouting of high-performance archers from a selected fitness and motor skill performance parameters. Sci. Sport. (2019). https://doi.org/10.1016/j.scispo.2019.02.006

2. H. Azahari, H. Juahir, M.R. Abdullah, R.M. Musa, V. Eswaramoorthi, N. Alias, S.M. Mat-Rashid, N.A. Kosni, A.B.H.M. Maliki, N.B. Raj, A multivariate analysis of cardiopulmonary parameters in archery performance. Hum. Mov. **19**, 35–41 (2019). https://doi.org/10.5114/hm.2018.77322

3. R.M. Musa, Z. Taha, A.P.P. Abdul Majeed, M.R. Abdullah, *Machine Learning in Sports: Identifying Potential Archers* (Springer Singapore, Singapore, 2019). https://doi.org/10.1007/978-981-13-2592-2

4. P.K. Suppiah, R. Muazu Musa, T. Wong, K. Kiet, M.R. Abdullah, Sensitivity prediction analysis of the contribution of physical fitness variables on Terengganu Malaysian Youth Archers' shooting scores. Int. J. Pharm. Sci. Rev. Res. **43**, 133–139 (2017)

5. M.N. Jawis, R. Singh, H.J. Singh, M.N. Yassin, Anthropometric and physiological profiles of sepak takraw players. Br. J. Sports Med. **39**, 825–829 (2005). https://doi.org/10.1136/bjsm.2004.016915

6. A. Rashid Aziz, E. Teo, B. Tan, T. Kong Oiuan, Sepak takraw : a descriptive analysis of heart rate and blood lactate response and physiological profiles of elite players. Int. J. Appl. Sport. Sci. **15**, 1–10 (2003)

7. N. Kosni, M. Abdullah, H. Juahir, R.M. Musa, A.B.H.M. Maliki, S.M. Mat-Rasid, A. Adnan, N. Alias, V. Eswaramoorthi, Determination association of anthropometric and performance ability in Sepak Takraw youth athlete using unsupervised multivariate. J. Fundam. Appl. Sci. **9**, 505 (2018). https://doi.org/10.4314/jfas.v9i2s.33

8. N.A. Kosni, M.R. Abdullah, M.N. Nazarudin, R.M. Musa, A.B.H.M. Maliki, A. Adnan, S.M. Mat-Rasid, H. Juahir, A comparative analysis of juggling skill between Sepak Raga and Bulu Ayam. J. Fundam. Appl. Sci. **10**, 857–868 (2018). https://dx.doi.org/10.4314/jfas.v10i1s.35

9. Y. Kubo, K. Nakazato, K. Koyama, Y. Tahara, A. Funaki, K. Hiranuma, The relation between hamstring strain injury and physical characteristics of Japanese collegiate Sepak Takraw Players. Int. J. Sports Med. **37**, 986–991 (2016). https://doi.org/10.1055/s-0042-114700

10. I. Hasanuddin, Z. Taha, N. Yusoff, N. Ahmad, R.A.R. Ghazilla, H. Usman, T.M.Y.S. Tuan Ya, Investigation of the head impact power of a Sepak Takraw ball on Sepak Takraw Players. Malays. J. Mov. Health Exerc. **4**(2), (2015). https://doi.org/10.15282/mohe.v4i2.21

11. J. Bangsbo, *The Physiology of Soccer*. https://link.springer.com/article/10.2165/00007256-200535060-00004 (1994)

12. G. Ziv, R. Lidor, Physical attributes, physiological characteristics, on-court performances and nutritional strategies of female and male basketball players. https://link.springer.com/article/10.2165/00007256-200939070-00003 (2009). https://doi.org/10.2165/00007256-200939070-00003

13. Z. Taha, R.M. Musa, P.P.A. Abdul Majeed, M.M. Alim, M.R. Abdullah, The identification of high potential archers based on fitness and motor ability variables: a support vector machine approach. Hum. Mov. Sci. **57**, 184–193 (2018). https://doi.org/10.1016/j.humov.2017.12.008

14. S.W. Chang, M.R. Abdullah, A.P.P.A. Abdul Majeed, A.F. Ab. Nasir, Z. Taha, R. Muazu Musa, A machine learning approach of predicting high potential archers by means of physical fitness indicators. PLoS One. **14**, e0209638 (2019). https://doi.org/10.1371/journal.pone.0209638

15. Z. Taha, R.M. Musa, A.P.P.A. Abdul Majeed, M.R. Abdullah, M.A. Abdullah, M.H.A. Hassan, Z. Khalil, The employment of support vector machine to classify high and low performance archers based on bio-physiological variables, in *IOP Conference Series Materials Science Engineering* 342, 012020 (2018). https://doi.org/10.1088/1757-899X/342/1/012020

Chapter 6
Relationship Between Psycho-Maturity and Performance of Sepak Takraw

6.1 Overview

Sepak takraw is regarded as a team sport that harnesses several skills and performance factors. The sport combined ball controlling skills such as kicking and juggling, agility and acrobatic ability as well as a greater level of reflexes. Thus, the sport can be considered as multitasking in nature in which the players are expected to anticipate sudden attacks from the opponent and to respond effectively within a short period of time. The aforementioned elements could be effectively managed in the event that the players are psychologically ready and mentally tough to endure the ever-demanding nature of the sport during the competition to ensure success [1]. The application of psychological strategies in this sport might enable the players to maintain the tempo and continue to attack as well as defend irrespective of the match status. The possession of appropriate psychological traits might give a player the advantage to force an opponent in making an unnecessary error at a highly competitive level.

There are a significant number of evidence supporting the importance of psychological traits in the sporting dimension [2, 3]. The psychological traits might involve a wide range of psychological elements that could be applied to determine the mental toughness as well as the psychological readiness of athletes. A study of psychological readiness is often carried out to examine the influence of such elements on the performance of athletes. Although it could be inferred that the major elements contributing to the performance of players in sepak takraw could be attributed to the physical, physiological as well as anthropometrics variables, however, a number of studies have stressed the importance of psychological traits in maintaining achievement and coping a variety of adversity that might arise during competition [4–7]. Moreover, it was reported that the possession of appropriate physical fitness, anthropometrics as well as physiological elements might not necessarily ensure success if an athlete is unable to set goals of maintaining the desired performance which is often attributed to the lack of suitable psychological traits [8].

© The Author(s), under exclusive license to Springer Nature Singapore Pte Ltd. 2020 49
R. Muazu Musa et al., *Machine Learning in Team Sports*,
SpringerBriefs in Applied Sciences and Technology,
https://doi.org/10.1007/978-981-15-3219-1_6

Maturity level coupled with psychological elements is reported to be some of the basic elements the coaches tend to identify in athletes for selection since the performance of athletes is often measured from a diverse range of performance parameters that are not only limited to physical fitness and physiological factors [9]. Growth and maturity are two important variables influencing performance in sports. Whilst growth refers to the quantifiable changes with respect to size, physical appearance as well as body composition in conjunction with several systems of the body, the term maturity reflects the variables inherent in the bodily systems with regard to the timing as well as temporal transformation [10]. It is worth to mention that these processes have been widely acknowledged by several researchers as important elements that could determine performance in numerous sports [11, 12]. Nonetheless, despite the reported importance of psychological elements as well as maturity status in the sporting domain, no studies have sought to determine the relationship of these variables towards the performance of sepak takraw game. Therefore, the present study is aimed at investigating the correlation of these variables to the successful performance of sepak takraw game and consequently identifies performance levels of youth sepak takraw players with regard to psycho-maturity variables, namely maturity status, self-talk, activation, imagery, emotion control, automaticity as well as goal setting.

Psycho-maturity evaluation: To evaluate the psychological traits of the players' understudy, the test of performance strategies–competition scale (TOPS-SC) was utilised. The Malaysian version of the TOPS-SC proposed by the previous researchers which measured seven basic psychological elements, namely maturity status, self-talk, activation, imagery, emotion control, automaticity and goal setting, was adopted in the present study [13]. Although it is worth to note that the original version of TOPS-SC measured numerous psychological traits, however, some of those traits were found to be insignificant with regard to the Malaysian athletes as reported in the aforementioned study as such only seven of the constructs are considered in the present study. The maturity assessment scale proposed by the preceding researcher was used to determine the level of maturity of the players in this investigation [14]. Moreover, the sepak takraw performance ability of the players was determined prior to the assessment of the above-mentioned variables in the study.

6.2   Clustering

The performance of the players from the measured variables, i.e. maturity index, self-talk, activation, imagery, emotion control, automaticity, as well as goal setting, was used as the input parameters for clustering. The Louvain clustering algorithm was used to cluster the players based on their performances in the assessed parameters. The algorithm is primarily employed to determine the performance class of the players to enable the allotment of group membership from the study variables.

6.3   Classification

In this investigation, the random forest model is utilised in order to determine the classification of the high, medium or low potential sepak takraw players from the extracted significant features. Two models are developed and compared, i.e. one (baseline model) that utilises two (2) and five (5) for the depth of the forest and the number of trees, respectively, against a random forest model that utilises optimised hyperparameters. A departure from the previous chapters, the hyperparameter optimisation of the model, namely the depth of the forest as well as the number of trees, is attained via validation curves. The ranges evaluated for the number of trees vary from 5 to 250 with an increment of 5 for each iteration, whilst for the depth of the forest, it is increased by 4 from 2 to 100. The data is demarcated into train and test dataset with a ratio of approximately 70:30, i.e. 51:23. From the training dataset, a fivefold cross-validation technique is employed to ascertain the suitable values for the aforesaid hyperparameters. The selection of the hyperparameter will be determined by observing the highest value of the classification accuracy from the cross-validation of the model that could also yield an accurate accuracy of the training set. The efficacy of the models developed is evaluated by evaluating the classification accuracy of both the training and the testing sets. Python's scikit-learn library through the Spyder platform was used in this study.

6.4   Results and Discussion

Figure 6.1 highlights the classes formed by the Louvain clustering algorithm based on the variables evaluated in the current investigation. It could be seen from the figure that three (3) groups are established with regard to the similarities of the performance of the players from the variables assessed. A clear allotment of group memberships could be observed from the figure which aids the assignation of group tag to each class, namely high psycho-matured (HPM) players, medium psycho-matured (MPM) players and low psycho-matured (LPM) players, respectively.

Figure 6.2 displays the performance variations of the players in accordance with the assessed variables in the study. The boxplots exhibited the mean performance of each group from the variables. It could be witnessed that the mean differences between the HMP and the MPM classes are higher from the LMP class. This evidence postulated that high and successful performance in the sepak takraw game to certain extent depends on the maturity status as well as some psychological skills.

It could be seen from the validation curve for the depth of the forest depicted in Fig. 6.3 that its optimised value is 26. Conversely, the optimised value for the number of trees is 50 (Fig. 6.4). The classification accuracies obtained for both the train and test datasets from the baseline model are 82.35% and 56.52%, respectively, whilst the classification accuracies for both the training and testing datasets obtained via the optimised model are 100 and 82.61% as shown in the bar chart illustrated in Fig. 6.5.

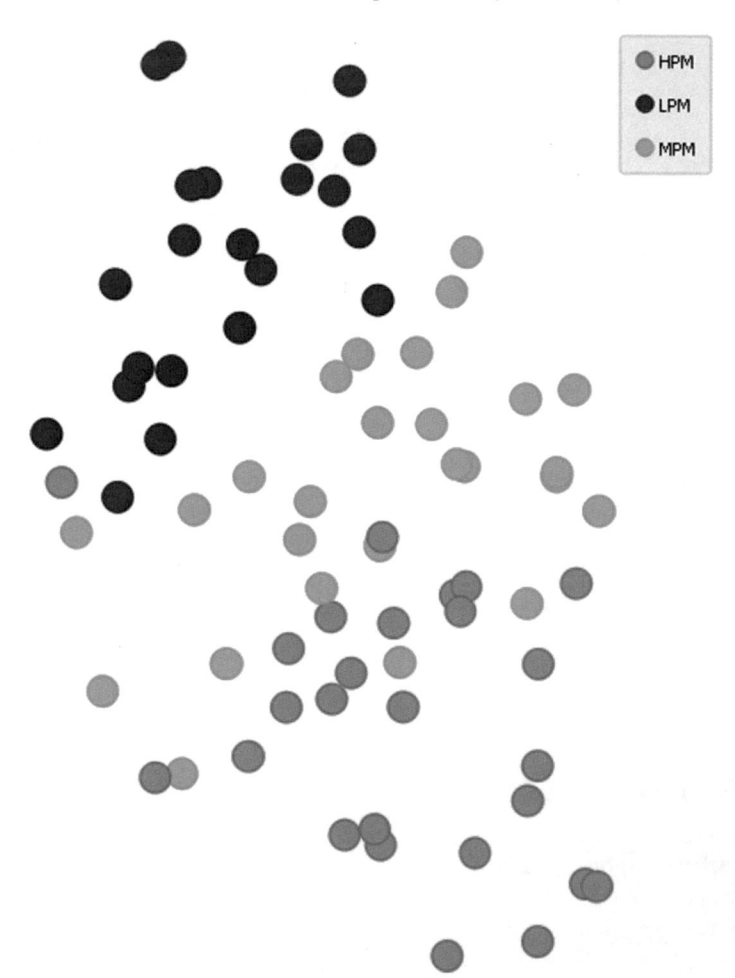

Fig. 6.1 Group membership assigned by the Louvain clustering algorithm

The confusion matrix of the train set as well as the test set is shown in Figs. 6.6 and 6.7, respectively. It could be seen that there is a slight misclassification transpired in Fig. 6.7, in which there are HPM players that are classified as MPM (2) and LPM (1). Also, it is worth noting that one of the MPM players is misclassified as an HPM player.

6.5   Summary

In the present chapter, the association between performance in sepak takraw and psycho-maturity variables is evaluated. It is apparent from the findings of the study

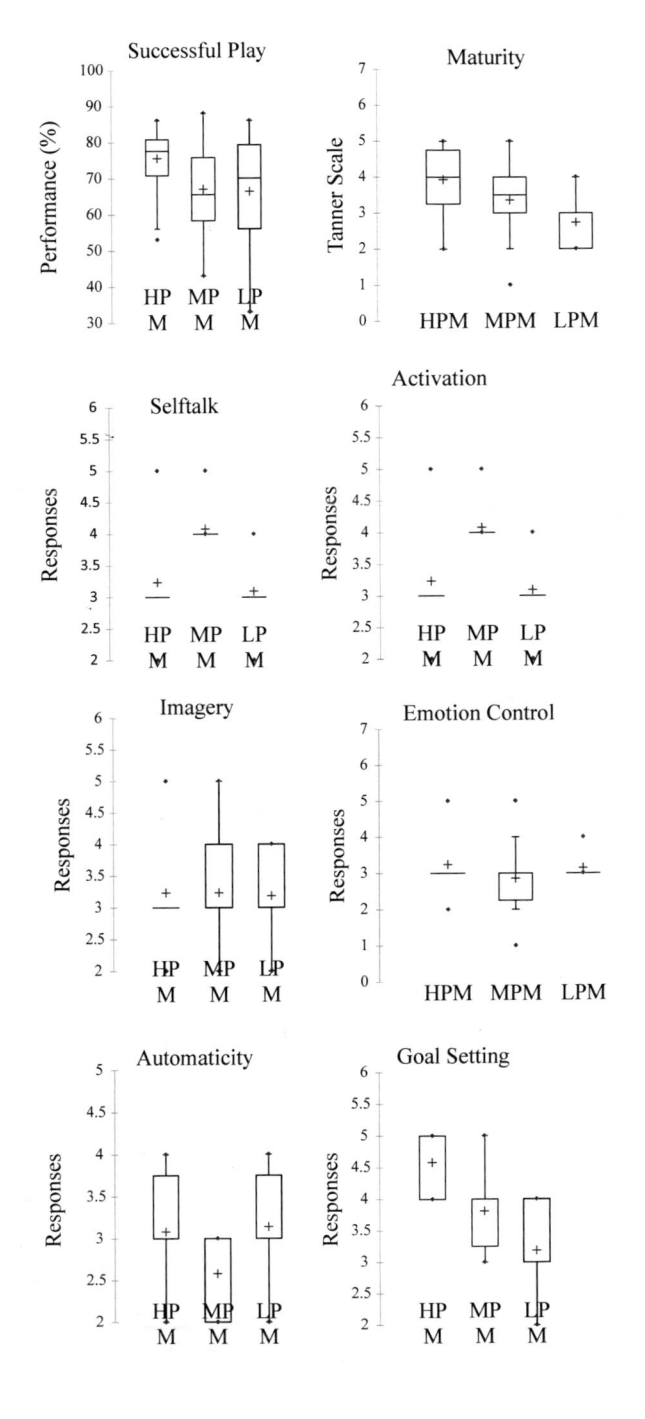

Fig. 6.2 Performance variations of the players based on the psycho-maturity variables

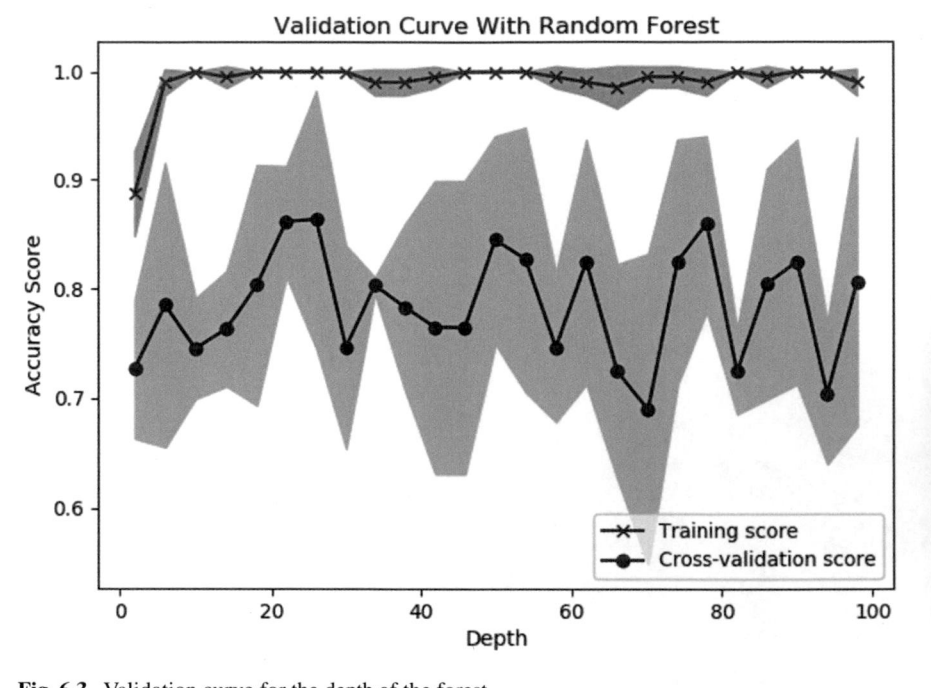

Fig. 6.3 Validation curve for the depth of the forest

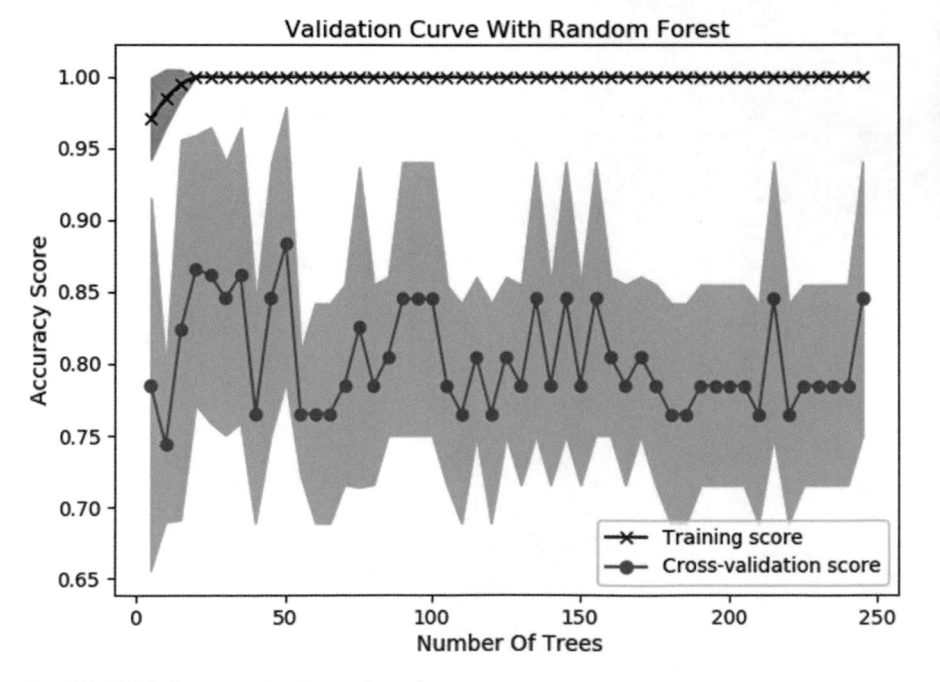

Fig. 6.4 Validation curve for the number of trees

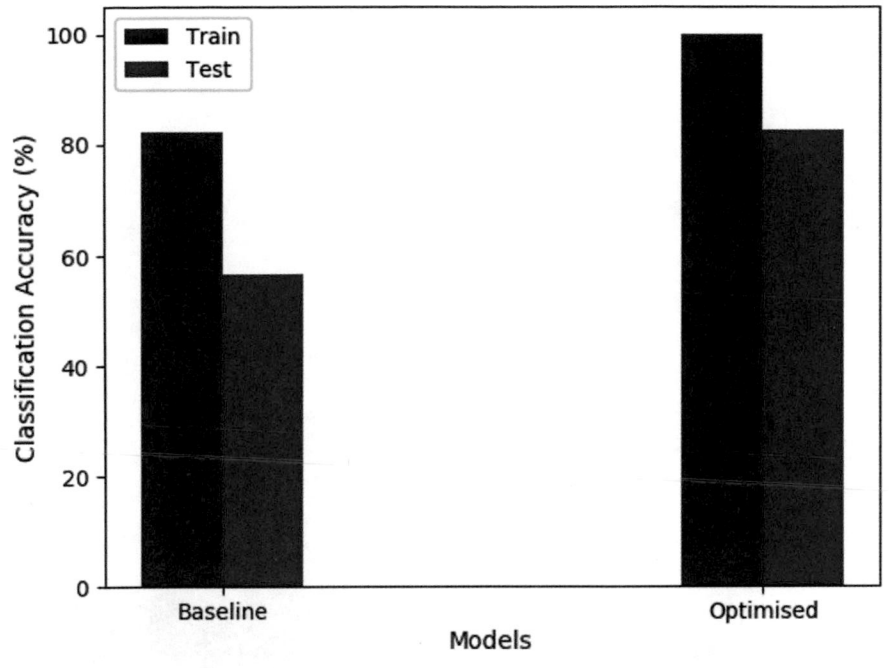

Fig. 6.5 Classification accuracy comparison between the baseline and optimised random forest model

that high performance in the sport could be reliant upon the attainment of psycho-maturity variables, specifically maturity status, self-talk, activation, imagery, emotion control, automaticity as well as goal setting. The nature of the sport as team-based highlights that the players are expected to possess some degree of maturity as well as a number of psychological skills. These elements are shown to be essential in assisting the players to cope with the psychological demands of the sport during both the game and training. Similarly, the results from the present investigation have demonstrated that a reasonably good classification accuracy could be attained by means of optimising the hyperparameters of the random forest classifier.

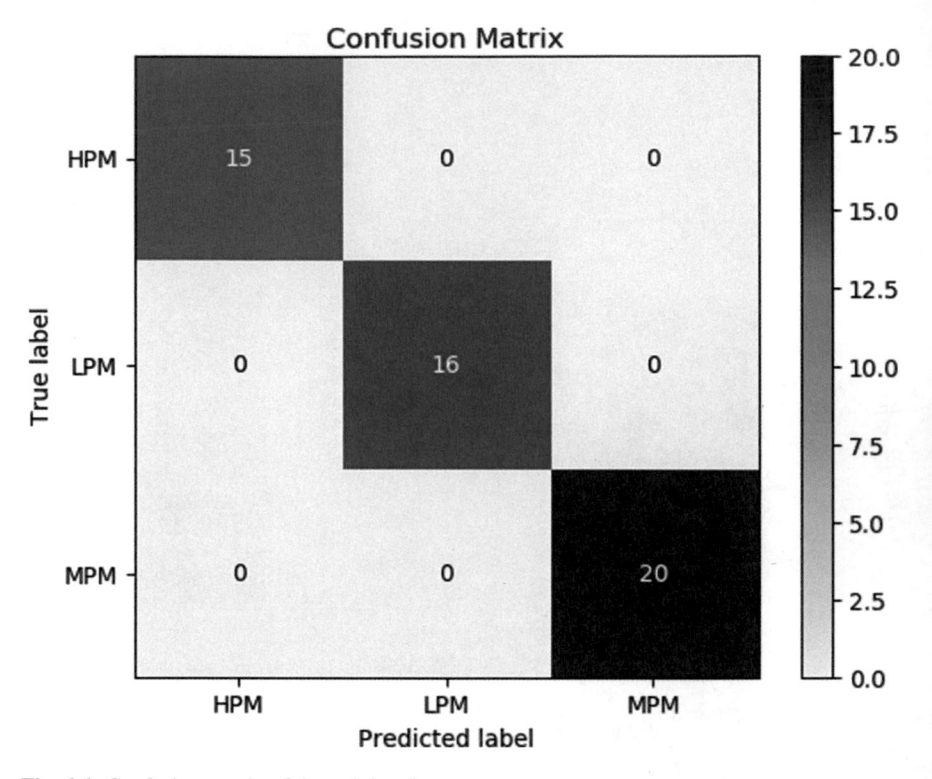

Fig. 6.6 Confusion matrix of the training data

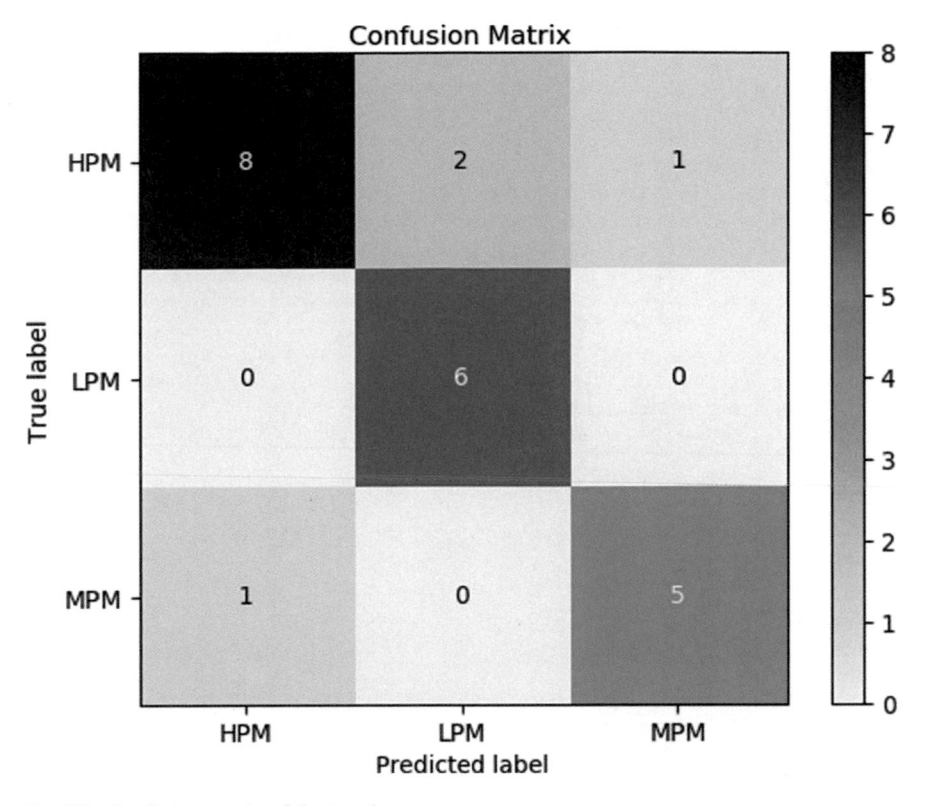

Fig. 6.7 Confusion matrix of the test data

References

1. M. Razali, R. Muazu, A. Maliki, N. Azura, P. Suppiah, Role of psychological factors on the performance of elite soccer players. J. Phys. Educ. Sport **16**, 170–176 (2016). https://doi.org/10.7752/jpes.2016.01027
2. J. Cameron, J. Cameron, L. Dithurbide, R. Lalonde, Personality traits and stereotypes associated with ice hockey positions. J. Sport Behav. **35**, 109 (2012)
3. M.K. Taylor, D. Gould, C. Rolo, Performance strategies of US Olympians in practice and competition. High Abil. Stud. **19**, 19–36 (2008). https://doi.org/10.1080/13598130801980281
4. J.N. Gilbert, W. Gilbert, C. Morawski, Coaching strategies for helping adolescent athletes cope with stress. J. Phys. Educ. Recreat. Danc. **78**, 13–24 (2007)
5. Z. Taha, R.M. Musa, A.P.P. Abdul Majeed, M.R. Abdullah, M.A. Zakaria, M.M. Alim, J.A.M. Jizat, M.F. Ibrahim, The identification of high potential archers based on relative psychological coping skills variables: a support vector machine approach, in *IOP Conference Series Materials Science Engineering*, vol. 319, 012027 (2018). https://doi.org/10.1088/1757-899X/319/1/012027
6. P. Fuster-Parra, A. García-Mas, F.J. Ponseti, F.M. Leo, Team performance and collective efficacy in the dynamic psychology of competitive team: A Bayesian network analysis. Hum. Mov. Sci. **40**, 98–118 (2015). https://doi.org/10.1016/j.humov.2014.12.005
7. R.E. Smith, D.S. Christensen, Psychological skills as predictors of performance and survival in professional baseball. J. Sport Exerc. Psychol. **17**, 399–415 (1995)

8. K. Kreiner-Phillips, T. Orlick, Winning after winning: the psychology of ongoing excellence. Sport Psychol. **7**, 31–48 (2016). https://doi.org/10.1123/tsp.7.1.31
9. N. Matos, R.J. Winsley, Trainability of young athletes and overtraining. J. Sports Sci. Med. **6**, 353 (2007)
10. R.M. Malina, Anthropometric correlates of strength end motor performance. Exerc. Sport Sci. Rev. **3**, 249–274 (1975)
11. F.V. Bastos, R.V. Hegg, The relationship of chronological age, body build, and sexual maturation to handgrip strength in schoolboys ages 10 to 17 years. Day, JAP. Perspect. kinanthropometry. Champaign Hum. Kinet., 45–49 (1986)
12. D.A. Santos, J.A. Dawson, C.N. Matias, P.M. Rocha, C.S. Minderico, D.B. Allison, L.B. Sardinha, A.M. Silva, Reference values for body composition and anthropometric measurements in athletes. PLoS ONE **9**, e97846 (2014)
13. M.R. Abdullah, N.A. Kosni, V. Eswaramoorthi, A.B.H.M. Maliki, R.M. Musa, Reliability of test of performance strategies-competition scale (TOPS-CS) among youth athletes: a preliminary study in Malaysia. Man India **96**, 5199–5207 (2016)
14. J.M. Tanner, Growth and maturation during adolescence. Nutr. Rev. **39**, 43–55 (1981)

Chapter 7
Concluding Remarks

A team sport is commonly regarded as any sporting activity that involves the harmonious efforts of the players to work together in the aspects not limited to communication, decision-making, conflict management, goal setting as well as fighting spirits. These combined efforts are essentially carried out in an atmosphere of support and trust with a specific goal of outperforming the opposing team. These elements often involve teammates facilitating the movement of the ball or other similar objects in accordance with set rules for the purpose of victory. Although there exist variations in the nature of team sports, nonetheless, it is worth noting that a number of performance-related parameters might be required for effective delivery of performance across many team sports. The nature of team sports showcases a multitude of challenges as well as a satisfaction to the players involved. Thus, the assimilation of both talents and physical prowess is needed from each team member in order to ensure success during competitions.

The present brief measures the associations of multiple performance-related parameters that contribute towards the delivery of performance in team-based sports, specifically beach soccer, as well as sepak takraw sports. In the beach soccer sport, key performance indicators that could define beach soccer performance are identified. The identified performance indicators are then investigated in their capacity to distinguish between winning and losing performances during the game. On the other hand, various human-related performance parameters ranging from anthropometry, physical fitness as well as psycho-maturity are assessed in an attempt to ascertain the possible relationship between these parameters towards the successful execution of performance in sepak takraw. Consequently, talented players are identified exclusively from their performances in the investigated human parameters through the employment of assorted machine learning-based algorithms.

It was demonstrated from the findings of the present brief that a number of performance indicators are essential in discriminating between losing and winning performances in beach soccer sport. The performance indicators, namely pass back third, shot back third, interception, turnover, goals scored in the second period, goals scored in the third period, as well as complete saves by the keeper, could influence team performance. These performance indicators can be categorised as technical as well as tactical which covered information on the defensive and offensive style of play of a given team during the competition. This investigation is exclusively geared towards the provision of relevant data that could be used to establish the normative profile of successful and unsuccessful performance in the Asian elite beach soccer tournaments. As such, the identified performance indicators herein could be important to the team coaches, managers as well as performance analysts in designing strategies that might bring about victory during competitions in relation to the technical and tactical performance indicators determined in this study.

Furthermore, the findings from the current brief illustrated that successful performance in the sport of sepak takraw is reliant upon several human-related performance markers. These performances of human factors are subcategorised into anthropometry, physical fitness as well as psycho-maturity. It has been revealed from the investigation that anthropometric indexes constituting standing height, sitting height, leg length, waist circumference, thigh circumference, calf circumference as well as four-site skinfold measurements (triceps, biceps, suprailiac and subscapular) do have an impact towards the performance of sepak takraw. The results highlighted that players with relatively bigger body physiques are shown to perform better compared to their counterparts. Conversely, some specific physical fitness parameters, particularly peak power velocity, explosive power, speed, dynamic as well as static balance, are shown to be indispensable in the determination of the successful performance of the sport. Further findings demonstrated that there exists an association between performance in the sport and some psycho-maturity elements, viz. maturity status, self-talk, activation, imagery, emotion control, automaticity as well as goal setting. This somewhat reflected that the players are required to possess a certain degree of maturity as well as psychological skills in order to cope with the competitive nature of the game.

Finally, the current brief employed the usage of a variety of machine learning-based algorithms to cater for the predetermined objectives of the current investigations. Although the application of machine learning in the sporting domain is relatively new, however, the ever-increasing accumulation of huge dataset in sports as well physical activity demonstrated that the current conventional method could no longer permit the extraction of information especially with regard to the data which is not linearly related to the sporting activity. Hence, the utilisation of non-conventional methods is, therefore, necessary in order to mitigate the problem of the nonlinearity nature of the dataset acquired in the sporting domain. It has been demonstrated from the findings of the present brief that the application of such machine learning methods is non-trivial in the identification of potential as well as high-performance athletes. The machine learning algorithms applied in the current investigations specifically artificial neural networks, k-nearest neighbour, support vector machine as well as random forest classifiers have further reiterated the capability of machine learning in assisting researchers, coaches, trainers, managers and other stakeholders in this domain to extracting useful information that could aid decision-making, training planning, talent identification and development as well as performance optimisation. It is recommended that the techniques explored in the present brief should be extended to other specific sports, games as well as physical activity.

Printed in the United States
By Bookmasters